SpringerBriefs in Probability and Mathematical Statistics

More information about this series at http://www.springer.com/series/14353

Kiyosi Itô

Poisson Point Processes and Their Application to Markov Processes

Foreword by Shinzo Watanabe and Ichiro Shigekawa

 Springer

Kiyosi Itô (1915–2008)
Kyoto University
Kyoto
Japan

Compiling Editors
Shinzo Watanabe
Kyoto University
Kyoto
Japan

Ichiro Shigekawa
Kyoto University
Kyoto
Japan

ISSN 2365-4333   ISSN 2365-4341   (electronic)
SpringerBriefs in Probability and Mathematical Statistics
ISBN 978-981-10-0271-7   ISBN 978-981-10-0272-4   (eBook)
DOI 10.1007/978-981-10-0272-4

Library of Congress Control Number: 2015959578

Printed on acid-free paper

This Springer imprint is published by SpringerNature
The registered company is Springer Science+Business Media Singapore Pte Ltd.

Foreword

Professor Kiyosi Itô had been leaving Japan and working regularly in foreign countries for several years including 1969. During the summer vacation of that year, he returned to Japan, to his home in Kyoto, and had a pleasant time meeting again and communicating with Japanese researchers and students. He gave then a series of lectures at Kyoto University and Osaka University. The present volume consists of original manuscripts of those lectures.

In the lectures, he treated a general theory of Poisson point processes and its applications to an extension problem (often called a boundary problem) of Markov processes. We know well that Itô applied the notion of Poisson point process to describe the sample functions of Lévy processes (additive processes, differential processes, or processes with independent increments) and thus established the Lévy–Itô decomposition theorem of Lévy processes.

As for the extension problem of Markov processes, he worked successfully with Henry McKean to describe the sample functions of Brownian motion on the half line [0, ∞) satisfying a Feller's boundary condition at 0. Then, he conceived a splendid idea that the collection of excursions from the boundary back to the boundary forms a Poisson point process taking values in the path space of excursions. It is a Poisson point process with values in an infinite dimensional function space, whereas in the case of the Lévy–Itô decomposition theorem for Lévy processes, the Poisson point process takes its value in a finite dimensional Euclidean space.

Although the application of Poisson point processes with values in the space of excursions treated in this volume is limited to the case of the boundary being a single point which is a discrete entrance point, this notion has proved to be a very useful and fundamental method in the boundary problem of Markov processes by many succeeding works including Itô's himself. We quote here Itô's recollection of this work from his "Foreword" in *Kiyosi Itô Selected Papers* (Springer, 1987): "After several years, it became my habit to observe even finite dimensional facts from the infinite dimensional view point. This habit led me to reduce the problem above to the Poisson point processes with values in the space of excursions."

Ichiro Shigekawa
Shinzo Watanabe

Preface

McKean and I determined all possible boundary conditions at 0 for the Brownian motion in $(0, \infty)$ and discussed the construction of the sample functions of the Markov processes corresponding to the boundary conditions [1]. The jumping-in measure k appearing in the boundary condition has to satisfy

$$\int_0^\infty (b \wedge 1)k(db) < \infty. \tag{1}$$

This conditions turns out to be

$$\int_0^\infty (s(b) \wedge 1)k(db) < \infty \tag{2}$$

for the diffusion in $(0, \infty)$ with the generator

$$\mathscr{G} = \frac{d}{dm}\frac{d}{ds} \tag{3}$$

if we have

$$s(0) > -\infty, \ m(0, 1) < \infty \text{ and } s(\infty) = \infty, \tag{4}$$

as we discussed in that paper. A few years ago, J. Lamperti raised the following question in connection with his work on branching processes.

What condition should the jumping-in measure k satisfy in case $m(0,1) = \infty$ in (4)?

By intuitive argument, I conjectured that the condition would be

$$\int_0^\infty E_b(1 - e^{-\sigma_0})k(db) < \infty, \quad \sigma_0 = \text{hitting time for } 0$$

or equivalently

$$\int_0^\infty \left(\int_0^b m(\xi, 1) \, ds(\xi) \wedge 1 \right) k(db) < \infty. \tag{5}$$

The purpose of this lecture is to solve this problem for the general Markov process with reasonable conditions by introducing the notion of the Poisson point process attached to the Markov process and to derive (2) and (5) as its special cases.

Let $Y_t(\omega)$ be a homogeneous Lévy process with paths increasing only with jumps.

Then,

$$E(e^{-\alpha Y_t}) = e^{-t \int_0^\infty (1 - e^{-\alpha u})n(du)} \tag{6}$$

where n is the Lévy measure of the process and

$$\int_0^\infty (u \wedge 1)n(du) < \infty. \tag{7}$$

Let D_ω be the discontinuity points of Y_t and consider the random set

$$G(\omega) = \{(t, Y_{t+}(\omega) - Y_{t-}(\omega)), t \in D_\omega\}.$$

This is a countable set in $T \times U$, $T = U = [0, \infty)$. It is well known that

(a) The number $\#(E \cap G)$ of points in $E \cap G$ is Poisson distributed with the mean:

$$\int_E dt \, n(du)$$

for every Borel set E in $T \times U$ (a random variable $\equiv \infty$ is regarded as Poisson distributed with mean $= \infty$) and

(b) $\#(E_i \cap G)$, $i = 1, 2, \ldots, n$ are independent for disjoint Borel sets E_i in $T \times U$.

These two conditions characterize the probability law of the random set G_ω.

Instead of considering the random set G_ω, we can consider the *point process* $X_\bullet(\omega)$ where $X_t(\omega)$ is defined only on D_ω and

$$X_t(\omega) = Y_{t+}(\omega) - Y_{t-}(\omega) \quad \text{for} \quad t \in D_\omega$$

for each ω. Then, G_ω is the graph of the path of $X.$. A point process in general is a random process whose sample function is defined only on a countable subset of the time interval depending on the sample.

The values of a point process need not be real. We can consider a point process whose values are taken from a general measurable space U. Let n be an arbitrary σ-finite measure on U. Then, a point process whose values are in U is called a *Poisson point process* with *characteristic measure* n, if its graph $G = G_X$ satisfies the conditions (a) and (b) mentioned above. We can define Poisson point processes in a qualitative way and derive (a) and (b) from the definition, as we shall do in this note.

In case the total measure $n(U)$ is finite, the domain of the definition of the sample function of the Poisson point process with characteristic measure n is a discrete set a.s. and its structure is simple. This was discussed by K. Matthes, J. Kersten, and P. Pranken [2–4]. It is a generalization of the point process arising from a compound Poisson process.

If $f : U \to U_1$ is measurable and if X is a Poisson point process: $T \to U$ with characteristic measure n, then the composition $f \cdot X$ is also a Poisson point process with characteristic measure nf^{-1}.

Let X_t be a Markov process on a locally compact metric space S and $a(\in S)$ be a fixed state. Let $A(t)$ be a local time process of X_t at a. Then, $A^{-1}(t)$ is a homogeneous Lévy process with increasing paths such that $P_a(A^{-1}(0) = 0) = 1$.

Let X_t^0 be a Markov process obtained by stopping X_t at the hitting time σ_a of X_t for a. σ_a is the same as the hitting time σ_a^0 of X_t^0 for a.

Let U be the space of all right continuous functions with left limits. We will define a point process $X: T \equiv [0, \infty) \to U$ by

$$D_{X_\omega}(= \text{the domain of } X_\omega)$$
$$= \text{the set of all discontinuity points of } A^{-1}(t)$$

and

$$X_{\omega,t}(s) = X(s + A^{-1}(t-)) \quad \text{if } s \leq A^{-1}(t+) - A^{-1}(t-)$$
$$= a \qquad\qquad\qquad\quad \text{if } s \geq A^{-1}(t+) - A^{-1}(t-)$$

for $t \in D_{X_\omega}$ (see the pictures in Sect. 2.2). We can use the strong Markov property of X_t to prove that X_ω is a Poisson point process: $T \to U$.

Let us introduce a function $e : U \to S$ by

$$e(u) = u(0).$$

Then, $e \cdot X$ is also a Poisson point process, and its characteristic measure is denoted by k and is called the *jumping-in measure* of X_t. Then, the characteristic measure n_X of X proves to be

$$n_X(V) = \int_S k(db) P_b(X_\bullet^0 \in V), \quad V \subset U$$

when $X_\bullet^0$ denotes the sample path of the stopped process X_t^0.

Let $h(u) = \inf \{t;\, u(t) = a\}$. Then, $h \cdot X$ is also a Poisson point process with characteristic measure $n_X \cdot h^{-1}$ and the jump part of $A^{-1}(t)$ is equal to

$$\sum_{\substack{s \in D_X \\ s \le t}} (h \cdot X)_s.$$

Using (7), we have

$$\int_0^\infty (t \wedge 1) n_X \cdot h^{-1}(dt) < \infty$$

i.e.,

$$\int_S k(db) E_b(\sigma_a^0 \wedge 1) < \infty.$$

Since the construction of a Poisson point process with a given characteristic measure is easy, we can discuss the construction of the Markov process X_t from its *stopped process*, its *jumping-in measure*, and its *stagnancy rate* (=the coefficient of t in the continuous part of $A^{-1}(t)$) if X_t has no continuous exit from a.

To discuss the case that a continuous exit from a is allowed, we will be faced with a more difficult problem. Roughly speaking, if we can determine all possible processes X_t with continuous exit only for their stopped process X_t^0 given (e.g., one-dimensional diffusion case), then we can determine all possible processes with both continuous exit and discontinuous exit. However, we will not discuss this problem in this note.

References

1. Itô, K., McKean, Jr., H.P.: Brownian motion on a half line. Ill. J. Math. **7**(2), 181–231 (1963)
2. Matthes, K.: Stationäre zufällige Punktfolgen. I, Jahresbericht d.D.M.V. **66**, 69–79 (1963)
3. Kersten, J., Matthes, K.: Stationäre zufällige Punktfolgen. II, Jahresbericht d.D.M.V. **66**, 106–118 (1964)
4. Pranken, P., Liemant, A., Matthes, K.: Stationäre zufällige Punktfolgen. III, Jahresbericht d.D.M.V. **67**, 183–202 (1965)

Contents

Chapter 1
Poisson Point Processes

1.1 Point Functions

Throughout this note we will use the following notations.

An interval of the type $[l, r)$, $-\infty < l < r \leq \infty$ is called a *time interval* and is denoted by $T, T_1, T_2, \ldots$. T is regarded as a measurable space associated with the topological σ-algebra on $\mathscr{T}$ on T. $\mathscr{T}_1, \mathscr{T}_2, \ldots$ are used respectively for those of $T_1, T_2, \ldots$.

$U, U_1, U_2, \ldots$ denote measurable spaces which are respectively associated with $\mathscr{U}, \mathscr{U}_1, \mathscr{U}_2, \ldots$. They are called *phase spaces*. In case $U_1 \subset U_2$, we assume $\mathscr{U}_1 = U_1 \cap \mathscr{U}_2$ (=trace σ-algebra of U_2 on U_1) unless the contrary is explicitly stated.

The *product space* $T \times U$ is regarded as a measurable space associated with the *product σ-algebra* $\mathscr{T} \times \mathscr{U}$.

Definition 1.1.1 A *point function* $p : T \to U$ is defined to be a map from a countable subset D_p into U. D_p is called the *domain* of p.

We admit an empty set for D_p. In this case p is called the *trivial point function*. If D_p has no accumulation point in T, p is called *discrete*.

Definition 1.1.2 The *graph* $G(p)$ of a point function $p : T \to U$ is defined to be

$$G(p) = \{(t, p(t)) : t \in D_p\}.$$

$G = G(p)$ is a countable subset of $T \times U$ such that every t-section of G, i.e., $\{u : (t, u) \in G\}$ is empty or a singleton. Conversely every countable subset of $T \times U$ with this property corresponds to a unique point function $: T \to U$.

For a point function $p : T \to U$ and a set $E \subset T \times U$ we write $N(p, E)$ for the number of points in $G(p) \cap E$.

Suppose that $T_1 \subset T_2$ and $U_1 \subset U_2$. Then every point function$: T_1 \to U_1$ is regarded as a point function $: T_2 \to U_2$.

© The Author(s) 2015   1
K. Itô, *Poisson Point Processes and Their Application to Markov Processes*,
SpringerBriefs in Probability and Mathematical Statistics,
DOI 10.1007/978-981-10-0272-4_1

Let $f : U \to U_1$. Then for a point function $p : T \to U$ we can define a point function $f \cdot p : T \to U_1$ by

$$D_{f \cdot p} = D_p, \quad (f \cdot p)(t) = f(p(t)) \quad \text{for} \quad t \in D_{f \cdot p}.$$

Let θ_s be a translation : $\theta_s t = t + s$ on $(-\infty, \infty)$. θ_s induces a set translation

$$\theta_s \cdot \mathscr{B} = \{t + s : t \in \mathscr{B}\}.$$

θ_s induces also a translation of a point function. Let p be a point function : $T \to U$. Then $\theta_s \cdot p$ is a point function : $\theta_s^{-1} T \to U$ defined by

$$D_{\theta_s \cdot p} = \theta_s^{-1} \cdot D_p, \quad (\theta_s \cdot p)(t) = p(t + s) \quad \text{for} \quad t \in D_{\theta_s \cdot p}.$$

Let p be a point function : $T \to U$ and $E \subset T \times U$. Then $G(p) \cap E$ corresponds to a unique point function which is called the *restriction* of p to E, $p|E$ in notation. For $T_1 \subset T$, the restriction $p|T_1 \times U$ is called the *domain restriction* of p to T_1, $p|_d T_1$ in notation. Similarly for $U_1 \subset U$, the restriction $p|T \times U_1$ is called the *range restriction of p to U_1*, $p|_r U_1$.

Let p_1 and p_2 be point functions : $T \to U$. If $G(p_1) \subset G(p_2)$, then p_2 is called an *extension* of p_1 and we write $p_2 \supset p_1$ to indicate this relation. If

$$p_1(t) = p_2(t) \quad \text{for} \quad t \in D_{p_1} \cap D_{p_2},$$

then p_1 and p_2 are called *consistent*. If $\{p_n\}$ is a countable family of point functions : $T \to U$ consistent with each other, then $\bigcup_n G(p_n)$ corresponds to a unique point function : $T \to U$ which is called the join of p_n, $\bigvee_n p_n$ in notation.

The space $\mathbb{P}(T, U)$ of all point functions: $T \to U$ is regarded as a measurable space associated with the σ-algebra $\mathscr{P}(T, U)$ generated by the sets

$$\{p \in \mathbb{P} : N(p, E) = k\}, \quad E \in \mathscr{T} \times \mathscr{U}, \quad k = 0, 1, 2, \ldots, \infty.$$

If $T_1 \subset T_2$ and $U_1 \subset U_2$, then $\mathbb{P}(T_1, U_1) \subset \mathbb{P}(T_2, U_2)$, namely $\mathbb{P}(T_1, U_1)$ is a subset of $\mathbb{P}(T_2, U_2)$ and

$$\mathscr{P}(T_1, U_1) = \mathbb{P}(T_1, U_1) \cap \mathscr{P}(T_2, U_2).$$

1.2 Point Processes

Let $(\Omega, \mathscr{B}, P)$ be a complete probability measure space. We write $\mathbb{P}$ and $\mathscr{P}$ for $\mathbb{P}(T, U)$ and $\mathscr{P}(T, U)$, respectively.

Definition 1.2.1 A function $X : \Omega \to \mathbb{P}$ measurable $\mathscr{B}|\mathscr{P}$ is called a point process (or a random point function) $: T \to U$ on $(\Omega, \mathscr{B}, P)$. The value of X at ω, X_ω in notation, is called a *sample point function* of X.

A point process $X : T \to U$ on $(\Omega, \mathscr{B}, P)$ is a random variable with values in $(\mathbb{P}, \mathscr{P})$. Therefore all notions concerning random valuables such as probability law, independence etc. are defined for point processes.

It follows from the definition that a map $X : \Omega \to \mathbb{P}$ is a point process if and only if $N(X, E)$ is measurable in ω for every E.

It is to be noted that if $T_1 \subset T$ and $U_1 \subset U$ then every point process $: T_1 \to U_1$ is also regarded as a point process $: T \to U$.

We can prove the following theorem by routine.

Theorem 1.2.2 *Two point processes X_1, $X_2 : T \to U$ have the same probability law, if we have*

$$P(N(X_1, E_i) = k_i, \ i = 1, 2, \ldots, n) = P(N(X_2, E_i) = k_i, \ i = 1, 2, \ldots, n)$$

for every n, every $\{k_i\}$ and every disjoint $\{E_i\}$.

The operations on point functions defined in Sect. 1.1 are also defined for point processes in the obvious sample-wise way. For example, the restriction $X|E$ is defined by

$$(X|E)_\omega = X_\omega|E.$$

It is obvious that if $E \in \mathscr{T} \times \mathscr{U}$ then $X|E$ is a point process $: T \to U$. Similarly for other operations.

A point process $X : T \to U$ is called *discrete* if X is a discrete point function a.s. $X : T \to U$ is called σ-*discrete* if we have an increasing sequence $\{U_n\} \subset \mathscr{U}$ such that $X|_r U_n$ is discrete for every n and that

$$X = X|_r \bigcup_n U_n \quad \text{a.s.}$$

$X : T \to U$ is called *differential* if $X|_d T_i, \ i = 1, 2, \ldots, n$ are *independent* for $\{T_i\}$ disjoint.

$X : T \to U$ is called *stationary* if $\theta_\tau(X|_d T_1)$ and $X|_d \theta_\tau^{-1} T_1$ have the same probability law as far as both T_1 and $\theta_\tau^{-1} T_1$ are included in T.

1.3 Poisson Point Processes

Definition 1.3.1 A point process $X : T = [0, \infty) \to U$ is called a *Poisson point process*, if it is σ-discrete, differential and stationary.

The name "Poisson point process" is justified by the following theorem.

Theorem 1.3.2 *Let X be a Poisson point process. Then we have the following properties.*

(a) For $E \in \mathcal{T} \times \mathcal{U}$, $N(X, E)$ is Poisson distributed[1].
(b) For $E_1, E_2, \ldots, E_n \in \mathcal{T} \times \mathcal{U}$ disjoint, $N(X, E_i)$, $i = 1, 2, \ldots, n$ are independent.

Proof Since X is σ-discrete, we have an increasing sequence $\{U_n\} \subset \mathcal{U}$ such that $X_n = X|_r U_n$ is discrete and that $X = X|_r \bigcup_n U_n$ a.s. Therefore

$$P\left(N(X, E) = \lim_{n \to \infty} N(X_n, E) \right) = 1,$$

and so our theorem holds if it holds for X_n. Thus it is enough to discuss the case that X is discrete.

Write $E[t]$ for the set $\{(s, u) \in E : s < t\}$. Then $Y(t) = N(X, E[t])$, $t \in T$, is a stochastic process whose sample function increases only by jumps $= 1$ a.s. It is obvious that $Y(0) = 0$. Since X is differential, $Y(t)$ is an additive process. We will prove that

$$P(Y(t-) = Y(t+)) = 1 \quad \text{for each } t \quad (Y(0-) = 0).$$

Consider the process:

$$Z(t) = N(X, [0, t) \times U).$$

Since X is differential and stationary, $Z(t)$ is an homogeneous additive process with increasing sample functions:

$$\varphi(t) = E(e^{-Z(t)}).$$

Then

$$\varphi(t + s) = E(e^{-(Z(t+s)-Z(t))})E(e^{-Z(t)}) = \varphi(s)\varphi(t).$$

Since $0 < \varphi(t) \le 1$, $\varphi(t) = e^{-\alpha t}$ with $0 \le \alpha < \infty$. Thus

$$E(e^{-(Z(t+)-Z(t-))}) = \lim_{\substack{t_1 \uparrow t \\ t_2 \downarrow t}} E(e^{-(Z(t_2)-Z(t_1))})$$

$$= \lim_{\substack{t_1 \uparrow t \\ t_2 \downarrow t}} \varphi(t_2 - t_1) = 1.$$

Therefore

$$P(Z(t+) - Z(t-) = 0) = 1.$$

Since $0 \le Y(t+) - Y(t-) \le Z(t+) - Z(t-)$ is obvious, we have $P(Y(t+) - Y(t-) = 0) = 1$. Thus $Y(t)$ is an additive process with no fixed discontinuities such

---

[1]The random variable $\equiv \infty$ is regarded to be Poisson distributed with mean ∞.

that its sample function increases only by jump $= 1$ a.s. and that $Y(0) = 0$. Therefore $Y(t)$ is Poisson distributed for each t. Since $N(X, E) = \lim_{t \to \infty} Y(t)$, $N(X, E)$ is also Poisson distributed. This proves (a).

To prove (b), consider the stochastic processes

$$Y_i(t) = N(X, E_i[t]), \quad i = 1, 2, \ldots, n,$$

$$Y(t) = \sum_{i=1}^{n} i Y_i(t).$$

Since X is differential, $Y(t)$ is an additive process. Since each $Y_i(t)$ is continuous in probability as proved above, $Y(t)$ is also continuous in probability. Since $Y_i(t)$ increases only by jump $= 1$, $Y_i(t)$ is the number of jumps $= i$ of the sample function of Y before t. Therefore $Y_i(t)$, $i = 1, 2, \ldots, n$ are independent (special case of the Lévy decomposition theorem [1, 2]). Letting $t \uparrow \infty$, we obtain (b). $\square$

Now we will investigate the structure of Poisson point processes. Let $X : T \to U$ be a Poisson point process. Then $N(X, E)$ is Poisson distributed. Set

$$m(E) = E(N(X, E)), \quad E \in \mathscr{T} \times \mathscr{U}.$$

Since $N(X, E)$ is a measure in E for every ω, $m(E)$ is also a measure. Let $\{U_n\} \subset \mathscr{U}$ be a sequence in the definition of the σ-discreteness of X. Then for $V \in \mathscr{U}$, $V \subset U_n$ for some n, we have

$$m([t_1, t_2) \times V) = m([t_1 + s, t_2 + s) \times V),$$
$$m([t_1, t_2) \times V) + m([t_2, t_3) \times V) = m([t_1, t_3) \times V)$$

and so

$$m([t_1, t_2) \times V) = (t_2 - t_1)n(V)$$

where

$$n(V) = m([0, 1) \times V).$$

It is obvious that n is a σ-finite measure on U and that $m(dt\,du) = dt \cdot n(du)$.

Definition 1.3.3 The measure n is called the *characteristic measure* of X.

The following theorem that follows at once from Theorems 1.2.2 and 1.3.2 shows that the characteristic measure n_X of a Poisson point process characterizes the probability law P_X.

Theorem 1.3.4 *Let X_1, X_2 be two Poisson point processes. $P_{X_1} = P_{X_2}$ if and only if $n_{X_1} = n_{X_2}$.*

Let us prove the *existence theorem* for Poisson point processes.

Theorem 1.3.5 *For a σ-finite measure n on $(U, \mathcal{U})$ given, there exists a Poisson point process X with $n_X = n$.*

Proof Since n is σ-finite, we have a mutually disjoint sequence $\{U_h\} \subset \mathcal{U}$ such that $n(U_h) < \infty$ and that $U = \bigcup_h U_h$. Let m be the product measure of the Lebesgue measure on T and the measure n on U, i.e., $m(dt\, du) = dt\, n(du)$. Let $V_{kh} = [k-1, k) \times U_h$. Then we have

$$T \times U = \bigcup_{k,h} V_{kh} \quad \text{(disjoint union)}$$

and

$$m(V_{kh}) = n(U_h) < \infty.$$

Consider a system of independent random variables[2,3]

$$N_{kh}, \quad k, h = 1, 2, \ldots, \quad X_{kh\lambda}, \quad k, h, \lambda = 1, 2, \ldots$$

such that N_{kh} is Poisson distributed with mean $m(V_{kh}) = n(U_h)$ and that $X_{kh\lambda}$ is distributed as follows:

$$P(X_{kh\lambda} \in E) = m(V_{kh} \cap E)/m(V_{kh});$$

the existence of such a system is well-known.

Let $\pi_1 : T \times U \to T$ be the projection map. First we will prove that $\{\pi_1(X_{kh\lambda})\}_{k,h,\lambda}$ are all different a.s. Since $X_{kh\lambda} \in [k-1, k) \times U$ a.s. for every k, h, λ, it is enough to prove that

$$P(\pi_1(X_{kh\lambda}) = \pi_1(X_{kj\mu})) = 0 \quad \text{except for} \quad (k, h, \lambda) = (k, j, \mu).$$

But

$$P(\pi_1(X_{kh\lambda}) = \pi_1(X_{kj\mu}))$$

$$\leq \sum_{\sigma=1}^{s} P(X_{kh\lambda} \in [t_{\sigma-1}, t_\sigma) \times U_h, X_{kj\mu} \in [t_{\sigma-1}, t_\sigma) \times U_j) \quad \left(t_\sigma = k - 1 + \frac{\sigma}{s}\right)$$

$$= \sum_{\sigma=1}^{s} P(X_{kh\lambda} \in [t_{\sigma-1}, t_\sigma) \times U_h) P(X_{kj\mu} \in [t_{\sigma-1}, t_\sigma) \times U_j)$$

$$= \sum_{\sigma=1}^{s} (t_\sigma - t_{\sigma-1})^2 = \frac{1}{s} \longrightarrow 0 \quad (s \to \infty).$$

---

[2]With values in $\{0, 1, 2, \ldots\}$.
[3]With values in $T \times U (\mathcal{T} \times \mathcal{U})$.

Thus we have proved that $\{\pi_1(X_{kh\lambda})\}_{k,h,\lambda}$ are all different a.s. Therefore the set

$$G(\omega) = \{X_{kh\lambda} : \lambda = 1, 2, \ldots, \ N_{kh}(\omega) : k, h = 1, 2, \ldots\}$$

defines a point function X_ω depending on ω.

Now we will prove that X is a Poisson point process. First we will prove that $N(X, E)$, $E \in \mathcal{T} \times \mathcal{U}$ is Poisson distributed. Since $N(X, E)$ is σ-additive in E, we can assume with no loss of generality that E is included in some V_{kh}. For $t \in [0, 1)$ we have

$$\begin{aligned}
E[t^{N(X,E)}] &= E\left[t^{\sum_{\lambda=1}^{N_{kh}} 1_E(X_{kh\lambda})} \right] \\
&= \sum_{\nu=0}^{\infty} P(N_{kh} = \nu) E\left[\prod_{\lambda=1}^{\nu} t^{1_E(X_{kh\lambda})} \right] \\
&= \sum_{\nu=0}^{\infty} P(N_{kh} = \nu) \prod_{\lambda=1}^{\nu} E[t^{1_E(X_{kh\lambda})}] \\
&= \sum_{\nu=0}^{\infty} e^{-c} \frac{c^\nu}{\nu!} \left(t\frac{d}{c} + \left(1 - \frac{d}{c}\right) \right)^\nu \quad (c = m(V_{kh}), d = m(E)) \\
&= e^{-c} e^{c\left(t\frac{d}{c} + (1 - \frac{d}{c})\right)} = e^{d(t-1)}.
\end{aligned}$$

This proves that $N(X, E)$ is Poisson distributed with mean $d = m(E)$.

Next we will prove that if $E_\beta \in \mathcal{T} \times \mathcal{U}$, $\beta = 1, 2, \ldots, \alpha$, are disjoint, then $N(X, E_\beta)$, $\beta = 1, 2, \ldots, \alpha$ are independent, i.e.,

$$E\left(\prod_{\beta=1}^{\alpha} t_\beta^{N(X,E_\beta)} \right) = \prod_{\beta=1}^{\alpha} E(t_\beta^{N(X,E_\beta)}) \left(\equiv \prod_{\beta=1}^{\alpha} e^{m(E_\beta)(t_\beta - 1)} \right).$$

Since the σ-algebras $\mathcal{B}_{kh}$ generated by $\{N_{kh}, X_{kh1}, X_{kh2}, \ldots\}$, $k, h = 1, 2, \ldots$ are independent and $N(X, E)$ is σ-additive in E, we can assume that $\{E_\beta\}_\beta$ is included in some V_{kh}. By adjoining $E_0 - V_{kh} \quad \bigcup_\beta E_\beta$ if necessary, we can assume that

$$V_{kh} = \bigcup_{\beta=1}^{\alpha} E_\beta \quad \text{(disjoint union)}.$$

Write 1_β for the indicator of E_β and suppress (k, h) in N_{kh}, V_{kh} and $X_{kh\lambda}$ for typographical symplicity.

$$E\left[\prod_{\beta=1}^{\alpha} t_\beta^{N(X,E_\beta)} \right] = E\left[\prod_{\beta=1}^{\alpha} t_\beta^{\sum_{\lambda=1}^{N} 1_\beta(X_\lambda)} \right]$$

$$= \sum_\nu P(N = \nu) E\Big[\prod_{\beta=1}^{\alpha} t_\beta^{\sum_{\lambda=1}^{\nu} 1_\beta(X_\lambda)}\Big]$$

$$= \sum_\nu P(N = \nu) E\Big[\prod_{\lambda=1}^{\nu} \prod_{\beta=1}^{\alpha} t_\beta^{1_\beta(X_\lambda)}\Big]$$

$$= \sum_\nu P(N = \nu) \prod_{\lambda=1}^{\nu} E\Big[\prod_{\beta=1}^{\alpha} t_\beta^{1_\beta(X_\lambda)}\Big]$$

$$= \sum_\nu P(N = \nu) \prod_{\lambda=1}^{\nu} \sum_{\beta=1}^{\alpha} t_\beta P(X_\lambda \in E_\beta)$$

$$= \sum_\nu e^{-m(V)} \frac{m(V)^\nu}{\nu!} \Big(\sum_{\beta=1}^{\alpha} t_\beta \frac{m(E_\beta)}{m(V)}\Big)^\nu$$

$$= e^{-m(V)} e^{\sum_\beta t_\beta m(E_\beta)}$$

$$= e^{\sum_\beta (t_\beta - 1) m(E_\beta)} = \prod_\beta e^{(t_\beta - 1) m(E_\beta)}.$$

It is now easy to prove that X is a Poisson point process with $n_X = n$. □

1.4 The Structure of Poisson Point Processes (1) the Discrete Case

Theorem 1.4.1 *Let X be a Poisson point process : $T \equiv [0, \infty) \to U$. We assume that X is discrete, namely that $n_X(U) < \infty$. Let*

$$D_{X_\omega} = \{\tau_1(\omega) < \tau_2(\omega) < \tau_3(\omega) < \cdots\}$$

and

$$\xi_i(\omega) = X_\omega(\tau_i(\omega)), \quad i = 1, 2, \ldots.$$

Then

(a) $P(\tau_i - \tau_{i-1} > t) = e^{-t n_X(U)}$ *($\tau_0 \equiv 0$), $i = 1, 2, \ldots,$*
(b) $P(\xi_i \in V) = n_X(V)/n_X(U)$, * $V \in \mathcal{U}$, $i = 1, 2, \ldots,$*
(c) $\tau_1, \tau_2 - \tau_1, \tau_3 - \tau_2, \ldots, \xi_1, \xi_2, \ldots$ *are independent.*

Proof Let $\alpha_i \geq 0$ and $V_i \in \mathcal{U}$, $i = 1, 2, \ldots, k$. We use the notation

$$\phi_p(t) = \frac{[pt] + 1}{p},$$

where $[a]$ = greatest integer $\leq a$. Then

$$E\left[e^{-\sum_{i=1}^{k}\alpha_i\tau_i}, \ \xi_i \in V_i, \ i = 1, 2, \ldots, k\right]$$

$$= \lim_{p\to\infty} E\left[e^{-\sum_{i=1}^{k}\alpha_i\phi_p(\tau_i)}, \ \xi_i \in V_i, \ \tau_i - \tau_{i-1} > \frac{1}{p}, \ i = 1, 2, \ldots, k\right]$$

$$= \lim_{p\to\infty} \sum_{0\leq v_1 < v_2 < \cdots < v_k} e^{-\sum_{i=1}^{k}\alpha_i\frac{v_i}{p}} P\left(\xi_i \in V_i, \ \frac{v_i - 1}{p} \leq \tau_i < \frac{v_i}{p}, \ i = 1, 2, \ldots, k\right).$$

But

$$P\left(\xi_i \in V_i, \ \frac{v_i - 1}{p} \leq \tau_i < \frac{v_i}{p}, \ i = 1, 2, \ldots, k\right)$$

$$= P\left(N\left(X, \left[\frac{v_i - 1}{p}, \frac{v_i}{p}\right) \times V_i\right) = 1, \ N\left(X, \left[\frac{v_i - 1}{p}, \frac{v_i}{p}\right) \times (U - V_i)\right) = 0,\right.$$

$$\left. N\left(X, \left[\frac{v-1}{p}, \frac{v}{p}\right) \times U\right) = 0 \ \text{ for } \ v \neq v_1, \ldots, v_k, \ v \leq v_k\right)$$

$$= \prod_i e^{-\frac{1}{p}n_X(V_i)} \frac{1}{p} n_X(V_i) e^{-\frac{1}{p}n_X(U - V_i)} \times \prod_{\substack{v \neq v_1, \ldots, v_k \\ v \leq v_k}} e^{-\frac{1}{p}n_X(U)} \quad \text{(by Theorem 1.3.2)}$$

$$= e^{-\frac{v_k}{p}n_X(U)} \prod_{i=1}^{k} \frac{1}{p} n_X(V_i).$$

Therefore

$$E\left[e^{-\sum_{i=1}^{k}\alpha_i\tau_i}, \ \xi_i \in V_i, \ i = 1, 2, \ldots, k\right]$$

$$= \lim_{p\to\infty} \sum_{0\leq v_1 < \cdots < v_k} e^{-\sum_{i=1}^{k}\alpha_i\frac{v_i}{p}} e^{-\frac{v_k}{p}n_X(U)} \prod_{i=1}^{k} n_X(V_i)\left(\frac{1}{p}\right)^k$$

$$= \prod_{i=1}^{k} n_X(V_i) \iint_{0\leq t_1 < \cdots < t_k} \cdots \int e^{-\sum_{i=1}^{k}\alpha_i t_i} e^{-t_k n_X(U)} \, dt_1 \cdots dt_k.$$

Set $\sigma_i = \tau_i - \tau_{i-1}, i = 1, 2, \ldots$. Then we have

$$E\left(e^{-\sum_{i=1}^{k}\beta_i\sigma_i}, \ \xi_i \in V_i, \ i = 1, 2, \ldots, k\right) \tag{1.1}$$

$$= \left(\prod_{i=1}^{k} \frac{n_X(V_i)}{n_X(U)}\right)\left(\prod_{i=1}^{k} \int_0^\infty e^{-\beta_i s} e^{-s\,n_X(U)} n_X(U) \, ds\right);$$

in fact,

the left side of (1.1)

$$= E\left[e^{-\sum_{i=1}^{k} \beta_i(\tau_i - \tau_{i-1})}, \ \xi_i \in V_i, \ i = 1, 2, \ldots, k\right]$$

$$= E\left[e^{-\sum_{i=1}^{k-1}(\beta_i - \beta_{i+1})\tau_i - \beta_k \tau_k}, \ \xi_i \in V_i, \ i = 1, 2, \ldots, k\right]$$

$$= \prod_{i=1}^{k} n_X(V_i) \int \cdots \int_{0 \le t_1 < \cdots < t_k} e^{-\sum_{i=1}^{k-1}(\beta_i - \beta_{i+1})t_i - \beta_k t_k} e^{-t_k n_X(U)} \, dt_1 \cdots dt_k$$

$$= \prod_{i=1}^{k} n_X(V_i) \int \cdots \int_{0 \le t_1 < \cdots < t_k} e^{-\sum_{i=1}^{k} \beta_i(t_i - t_{i-1})} e^{-t_k n_X(U)} \, dt_1 \cdots dt_k$$

$$= \prod_{i=1}^{k} n_X(V_i) \int \cdots \int_{s_1, s_2, \ldots, s_k \ge 0} e^{-\sum_{i=1}^{k} \beta_i s_i} e^{-\left(\sum_{i=1}^{k} s_i\right) n_X(U)} \, ds_1 \cdots ds_k$$

$$= \text{the right side of (1.1)}.$$

Setting $V_{i_0} = V$, $V_i = U$ ($i \ne i_0$) and all $\beta_i = 0$ in (1.1), we have

$$P(\xi_{i_0} \in V) = \frac{n_X(V)}{n_X(U)}, \tag{1.2}$$

proving (b).

Setting $\beta_{i_0} = \beta$, $\beta_i = 0$ ($i \ne i_0$) and all $V_i = U$ in (1.1), we have

$$E(e^{-\beta \sigma_{i_0}}) = \int_0^\infty e^{-\beta s} e^{-s\,n_X(U)} n_X(U) \, ds, \tag{1.3}$$

proving (a).

Using (1.2) and (1.3) we can write (1.1) as follows;

$$E\left[e^{-\sum_{i=1}^{k} \beta_i \sigma_i}, \ \xi_i \in V_i, \ i = 1, 2, \ldots, k\right] = \prod_{i=1}^{k} P(\xi_i \in V_i) \prod_{i=1}^{k} E[e^{-\beta_i \sigma_i}],$$

which proves (c).   □

In view of this theorem we can give a new contruction of a Poisson point process X with $n_X = n$ for a given bounded measure n on U.

Theorem 1.4.2 *Let* $\sigma_1, \sigma_2, \ldots, \xi_1, \xi_2, \ldots$ *be independent such that*

$$P(\sigma_i > t) = e^{-t\,n(U)}$$

and that

$$P(\xi_i \in V) = n(V)/n(U);$$

the existence of such a family $\{\sigma_i, \xi_i\}_i$ is well-known.
    Now set

$$X_\omega(t) = \xi_i \quad for \quad t = \sigma_1 + \sigma_2 + \cdots + \sigma_i,$$
$$D_X = \{\sigma_1, \sigma_1 + \sigma_2, \sigma_1 + \sigma_2 + \sigma_3, \ldots\}.$$

Then X is a Poisson point process with $n_X = n$.

Proof It suffices by the definition of Poisson point processes to prove the following fact.
    Let

$$0 = t_0 < t_1 < t_2 < \cdots < t_p$$

and

$$U = \bigcup_{i=1}^{q} V_i \quad \text{(disjoint)}, \quad V_i \in \mathcal{U}.$$

Then

$$N(X, [t_{i-1}, t_i) \times V_j), \quad i = 1, 2, \ldots, p, \quad j = 1, 2, \ldots, q,$$

are independent and each one is Poisson distributed with mean $(t_i - t_{i-1})n(V_j)$.
    By our assumption the process

$$Y(t) = \begin{cases} 0, & t < \sigma_1 \\ k, & \sigma_1 + \sigma_2 + \cdots + \sigma_k \leq t < \sigma_1 + \sigma_2 + \cdots + \sigma_{k+1}, \quad k = 1, 2, 3, \ldots \end{cases}$$

is a Poisson process with parameter $= n(U)$ independent of the family $(\xi_1, \xi_2, \ldots)$.
    Take an arbitrary family of non-negative integers:

$$v_{ij}, \quad i = 1, 2, \ldots, p, \quad j = 1, 2, \ldots, q$$

and set

$$v_i = \sum_j v_{ij}, \quad \mu_i = \sum_{\alpha=1}^{i} v_\alpha.$$

Then

$$P(N(X, [t_{i-1}, t_i) \times V_j) = v_{ij}, \ 1 \leq i \leq p, \ 1 \leq j \leq q)$$
$$= P(Y(t_i) - Y(t_{i-1}) = v_i, \ \#_j(\xi_{\mu_{i-1}+1}, \xi_{\mu_{i-1}+2}, \ldots, \xi_{\mu_i}) = v_{ij},$$
$$1 \leq i \leq p, \ 1 \leq j \leq q),$$

where $\#_j(\xi_{\alpha+1}, \ldots, \xi_\beta)$ denotes the number of points in

$$\{\xi_{\alpha+1}, \ldots, \xi_\beta\} \cap V_j.$$

This is equal to

$$
\begin{aligned}
&P(Y(t_i) - Y(t_{i-1}) = v_i, \ 1 \le i \le p) \\
&\quad \times P(\#_j(\xi_{\mu_{i-1}+1}, \ldots, \xi_{\mu_i}) = v_{ij}, \ 1 \le i \le p, \ 1 \le j \le q) \\
&= \prod_{i=1}^{p} P(Y(t_i) - Y(t_{i-1}) = v_i) \\
&\quad \times \prod_{i=1}^{p} P(\#_j(\xi_{\mu_{i-1}+1}, \ldots, \xi_{\mu_i}) = v_{ij}, \ 1 \le j \le q) \\
&= \prod_{i=1}^{p} e^{-(t_i - t_{i-1})n(U)} \frac{((t_i - t_{i-1})n(U))^{v_i}}{v_i!} \\
&\quad \times \prod_{i=1}^{p} \frac{v_i!}{v_{i1}! \cdots v_{iq}!} \left(\frac{n(V_1)}{n(U)}\right)^{v_{i1}} \cdots \left(\frac{n(V_q)}{n(U)}\right)^{v_{iq}} \\
&= \prod_{i=1}^{p} \prod_{j=1}^{q} e^{-(t_i - t_{i-1})n(V_j)} \frac{((t_i - t_{i-1})n(V_j))^{v_{ij}}}{v_{ij}!}
\end{aligned}
$$

from which we can easily derive the fact mentioned above. $\square$

1.5 The Structure of Poisson Point Processes (2) the General Case

As an immediate consequence of the definition of Poisson point processes we have:

Theorem 1.5.1 *Let X be a Poisson point process $T \equiv [0, \infty) \to U$ and $V \in \mathscr{U}$. Then the range restriction $Y \equiv X|_r V$ is a Poisson point process $: T \to V$.*
If $n_X(V) < \infty$, then $n_Y(V) = n_X(V) < \infty$ and therefore discrete.

In view of this fact we have:

Theorem 1.5.2 *A general $(=\sigma$-discrete) Poisson point process is the join of an extending sequence of discrete Poisson point processes. The characteristic measure of the original process is the limit of the characteristic measures of the discrete Poisson point processes.*

Proof Let X be a σ-discrete Poisson point process $: T \to U$. Then we have

$$U_k \uparrow U, \quad U_k \in \mathscr{U}, \quad n(U_k) < \infty.$$

Let $X_k = X|_r U_k$. Then X_k is a discrete Poisson point process and X_{k+1} is an extension of X_k for $k = 1, 2, \ldots$.

We have obviously

$$X = \bigvee_k X_k, \quad \text{and} \quad n_X = \lim_{k \to \infty} n_{X_k};$$

completing the proof.   □

As an immediate consequence of Theorem 1.3.2 we have:

Theorem 1.5.3 *Let X be a general ($=\sigma$-discrete) Poisson point process : $T = [0, \infty) \to U$, and*

$$U = \bigcup_k U_k \quad (disjoint), \quad U_k \in \mathscr{U}, \quad n_X(U_k) < \infty.$$

Then $X_k \equiv X|_r U_k$, $k = 1, 2, \ldots$ are independent discrete Poisson point processes, and

$$X = \bigvee_k X_k.$$

In view of this fact we can give a new construction of a general Poisson point processes with any given σ-finite characteristic measure.

Theorem 1.5.4 *Let n be a σ-finite measure on U such that*

$$U = \bigcup_k U_k \quad (disjoint), \quad U_k \in \mathscr{U}, \quad n(U_k) < \infty.$$

Let X_k be a Poisson point process with characteristic measure $n_{X_k}(\cdot) \equiv n(\cdot \cap U_k)$ for $k = 1, 2, \ldots$ and suppose that $\{X_k\}_k$ are independent. (Such a sequence $\{X_k\}_k$ can be constructed by virtue of Theorem 1.4.2 and the existence theorem of the product measure.) Then $X \equiv \bigvee_k X_k$ is a Poisson point process with $n_X = n$.

1.6   Transformation of Poisson Point Processes

Let $X : T \to U$ be a Poisson point process and $f : U \to U_1$ be measureble $\mathscr{U}/\mathscr{U}_1$. Then the composition $f \circ X$ is a point process.

Theorem 1.6.1 *If $f \circ X$ is σ-discrete, then $f \circ X$ is a Poisson point process, and $n_{f \circ X} = n_X f^{-1}$.*

Proof It is obvious that $f \circ X$ is differential and stationary. Since $f \circ X$ is σ-discrete by the assumption, $f \circ X$ is a Poisson point process. Since the number of points in the set

$$\{t : t \in D_{f \circ X} \cap [0, 1), \ (f \circ X_\omega)(t) \in V_1\}, \quad V_1 \in \mathcal{U}_1$$

is the same as that in the set

$$\{t : t \in D_X \cap [0, 1), \ X_\omega(t) \in f^{-1}(V_1)\}, \quad V_1 \in \mathcal{U}_1,$$

we have

$$n_{f \circ X} = n_X f^{-1}.$$

$\square$

1.7 Summable Point Processes

Let X be a point process : $T = [l, r) \to [0, \infty)$.

Definition 1.7.1 X is called *summable* if

$$P\left(\sum_{l \leq t \leq s} X_t < \infty\right) = 1 \quad \text{for every } s \in T.$$

Suppose that X_t is a summable point process. Then

$$S_t = \sum_{l \leq s \leq t} X_s, \quad t \in T$$

is a stochastic process with values in the space of increasing non-negative right continuous functions and

$$D_X = \text{ the set of all discontinuity points of } S_t,$$
$$X_t = S_t - S_{t-} \quad \text{for t} \in D_X.$$

Definition 1.7.2 S_t is called the *integrated process* of X_t.

Theorem 1.7.3 *A Poisson point process* $X : T = [l, r) \to [0, \infty)$ *is summable if and only if its integrated process is a* homogeneous increasing Lévy process *(=a subordinator).*

Proof Obvious by the definitions. $\square$

1.8   The Strong Renewal Property of Poisson Point Processes

Let X be a Poisson point process : $T \to U$ on $(\Omega, \mathscr{B}, P)$. Recall that P is complete, namely that $\mathscr{B} = \mathscr{B}^P$. Write $\mathscr{B}_{st}$ for the P-completion of the σ-algebra $\mathscr{B}[X|_d[s, t)]$ and $\mathscr{B}_t$ for $\mathscr{B}_{0t}$. It is obvious that $\mathscr{B}_t$ increases with t. Since X is differential, $\mathscr{B}_t$ and $\mathscr{B}_{t\infty}$ are independent.

Theorem 1.8.1 $\mathscr{B}_t$ *is right continuous, i.e.,*

$$\mathscr{B}_t = \bigcap_{s>t} \mathscr{B}_s.$$

Proof Set $\mathscr{B}_{t+} = \bigcap_{s>t} \mathscr{B}_s$. It is enough to prove that every bounded $\mathscr{B}_{t+}$-measurable function f is $\mathscr{B}_t$-measurable. f is obviously $\mathscr{B}_{t+1}$-measurable. Therefore we can find a continuous function φ on $\{0, 1, 2, \ldots, \infty\}^k$, k being some finite integer, and $\{E_i, i = 1, 2, \ldots, k\} \subset \mathscr{T} \times \mathscr{U}$ such that

$$E_i \subset [0, t+1) \times U$$

and that

$$E|f - \varphi(N(X, E_1), \ldots, N(X, E_k))| < \epsilon. \tag{1.4}$$

Since X is σ-discrete, we have an increasing sequence $\{U_p\} \subset \mathscr{U}$ such that $n(U_p) < \infty$ and that $X = X|_r \bigcup_p U_p$ a.s. Then

$$N(X, E_i) = \lim_{p\to\infty} N(X, E_i \cap ([0, t+1) \times U_p)).$$

Since φ is continuous, we can assume that every E_i in (1.4) is included in $[0, t + 1) \times U_p$. Write $E^-(s)$ for $[0, s) \times U$ and $E^+(s) = [s, \infty) \times U$. Then

$$P(N(X, E_i) \neq N(X, E_i \cap E^-(t)) \mid N(X, E_i \cap E^+(t + \delta)))$$
$$\leq P(N(X, [t, t+\delta) \times U_p) \neq 0)$$
$$= 1 - e^{-\delta n(U_p)} \longrightarrow 0 \quad \text{as} \quad \delta \to 0.$$

Therefore we can assume that every E_i in (1.4) included either in $E^-(t)$ or in $E^+(t + \delta)$ for some $\delta > 0$ independent of i. Suppose that $E_1, \ldots, E_j \subset E^-(t)$ and $E_{j+1}, \ldots, E_k \subset E^+(t + \delta)$. Since φ is continuous, we have

$$E\left|\varphi(N(X, E_1), \ldots, N(X, E_k)) - \sum_{\alpha=1}^{\nu} g_\alpha h_\alpha\right| < \varepsilon, \tag{1.5}$$

where each g_α is a continuous function of $N(X, E_i)$, $i \le j$ and h_α is a continuous function of $N(X, E_i)$, $j < i \le k$. Then g_α is $\mathscr{B}_t$-measurable and h_α is $\mathscr{B}_{t+\delta,\infty}$-measurable. Since $\mathscr{B}_{t+} \subset \mathscr{B}_{t+\delta}$, $\mathscr{B}_{t+\delta,\infty}$ is independent of $\mathscr{B}_{t+}$. Therefore every h_α is independent of $\mathscr{B}_{t+}$. Then we have

$$2\epsilon > E\left|f - \sum_\alpha g_\alpha h_\alpha\right|$$
$$= E\left[E\left(\left|f - \sum_\alpha g_\alpha h_\alpha\right| \middle| \mathscr{B}_{t+}\right)\right]$$
$$\ge E\left|E\left(f - \sum_\alpha g_\alpha h_\alpha \middle| \mathscr{B}_{t+}\right)\right|$$
$$= E\left|f - \sum_\alpha g_\alpha E(h_\alpha)\right|$$

because f and g_α are $\mathscr{B}_{t+}$-measurable and h_α is independent of $\mathscr{B}_{t+}$. Since $\sum_\alpha g_\alpha E(h_\alpha)$ is $\mathscr{B}_t$-measurable, it is now easy to complete the proof of the $\mathscr{B}_t$-measurability of f.   □

Stopping times and stopped σ-algebras are defined as in the theory of Markov processes.

Definition 1.8.2 A random time $\sigma = \sigma(\omega) \ge 0$ is called a *stopping time* with respect to $\{\mathscr{B}_t\}_t$ if $(\sigma \le t) \in \mathscr{B}_t$ for every t.

The stopped σ-algebra $\mathscr{B}_\sigma$ by a stopping time σ is defined by

$$\mathscr{B}_\sigma = \{B \in \mathscr{B}_\infty : B \cap (\sigma \le t) \in \mathscr{B}_t \text{ for every } t\}.$$

Since X is differential and stationary, it is easy to verify the *renewal property* :

$$P(B \cap (\theta_t X \in M)) = P(B)P(X \in M), \quad B \in \mathscr{B}_t, \quad M \in \mathscr{P}$$

for every t fixed.

Let $\mathscr{T}^+$ be the topological σ-algebra on $T^+ \equiv (0, \infty)$ and $\mathscr{P}^+$ the σ-algebra generated by the sets

$$\{f \in \mathbb{P} : N(f, E) = k\}, \quad k = 0, 1, 2, \ldots, \quad ; \quad E \in \mathscr{T}^+ \times \mathscr{U}.$$

Theorem 1.8.3 (strong renewal property) *If σ is a stopping time with respect to $\{\mathscr{B}_t\}$ such that*
$$P(\sigma < \infty) = 1,$$

then

$$P(B \cap (\theta_\sigma X \in M)) = P(B)P(X \in M), \quad B \in \mathscr{B}_\sigma, \quad M \in \mathscr{P}^+. \tag{1.6}$$

Proof Since both sides are bounded measures in $M \in \mathscr{P}^+$, it is enough to prove this on a semi-multiplicative[4] class which generates the σ-algebra $\mathscr{P}^+$. Let $\mathscr{C}$ be the class of all subsets M of $\mathbb{P}$ of the form

$$M = \{f : N(f, [s_i, t_i) \times V_i) = k_i, \ i = 1, 2, \ldots, p\}$$

where $[s_i, t_i) \times V_i, i = 1, 2, \ldots, p$ are disjoint and

$$0 < s_i < t_i < \infty, \quad n(V_i) < \infty \quad \text{for} \quad i = 1, 2, \ldots, p.$$

Then $\mathscr{C}$ is a semi-multiplicative class which generates the σ-algebra $\mathscr{P}^+$. It is therefore enough to prove our identity for $M \in \mathscr{C}$. For this purpose it suffices to prove

$$E\left\{\prod_i \alpha_i^{N(\theta_\sigma X, [s_i, t_i) \times V_i)}, B\right\} = E\left\{\prod_i \alpha_i^{N(X, [s_i, t_i) \times V_i)}\right\} P(B)$$

for $0 \le \alpha_i \le 1$.

Since $N(f, E)$ is a measure in $E \in \mathscr{T} \times \mathscr{U}$, we have

$$E\left\{\prod_i \alpha_i^{N(\theta_\sigma X, [s_i, t_i) \times V_i)}, B\right\}$$

$$= E\left\{\prod_i \alpha_i^{N(X, [s_i + \sigma, t_i + \sigma) \times V_i)}, B\right\}$$

$$= \lim_{k \to \infty} \lim_{h \to \infty} E\left\{\prod_i \alpha_i^{N(X, [s_i - \frac{1}{k} + \frac{1}{h} + \sigma, t_i - \frac{1}{k} + \sigma) \times V_i)}, B\right\}$$

$$= \lim_{k \to \infty} \lim_{h \to \infty} \sum_{j=1}^{\infty} E\left\{\prod_i \alpha_i^{N(X, [s_i - \frac{1}{k} + \frac{1}{h} + \sigma, t_i - \frac{1}{k} + \sigma) \times V_i)}, B \cap \left(\frac{j-1}{h} \le \sigma < \frac{j}{h}\right)\right\}.$$

Taking k big enough, we get

$$s_i - \frac{1}{k} > 0, \quad i = 1, 2, \ldots, p.$$

Then

$$\left[s_i - \frac{1}{k} + \frac{j}{h}, \ t_i - \frac{1}{k} + \frac{j}{h}\right) \times V_i, \quad i = 1, 2, \ldots, p$$

are disjoint, because they are respectively equal to the sets

$$\theta_{-\frac{1}{k} + \frac{j}{h}} \cdot ([s_i, t_i) \times V_i), \quad i = 1, 2, \ldots, p.$$

---

[4]$\mathscr{C}$ is called *semi-multiplicative* if the intersection of two members in $\mathscr{C}$ is expressed as a finite disjoint union of members in $\mathscr{C}$.

Therefore

$$\left[s_i - \frac{1}{k} + \frac{j+1}{h}, \ t_i - \frac{1}{k} + \frac{j-1}{h} \right) \times V_i, \quad i = 1, 2, \ldots, p$$

are disjoint as well.

For big k, we have

$$\sum_{j=1}^{\infty} E\left\{ \prod_i \alpha_i^{N(X, [s_i - \frac{1}{k} + \frac{1}{h} + \sigma, \ t_i - \frac{1}{k} + \sigma) \times V_i)}, \ B \cap \left(\frac{j-1}{h} \le \sigma < \frac{j}{h} \right) \right\}$$

$$\ge \sum_j E\left\{ \prod_i \alpha_i^{N(X, [s_i - \frac{1}{k} + \frac{j+1}{h}, \ t_i - \frac{1}{k} + \frac{j-1}{h}) \times V_i)}, \ B \cap \left(\frac{j-1}{h} \le \sigma < \frac{j}{h} \right) \right\}$$

$$= \sum_j E\left\{ \prod_i \alpha_i^{N(X, [s_i - \frac{1}{k} + \frac{j+1}{h}, \ t_i - \frac{1}{k} + \frac{j-1}{h}) \times V_i)} \right\} P\left(B \cap \left(\frac{j-1}{h} \le \sigma < \frac{j}{h} \right) \right)$$

$$= \sum_j \prod_i e^{(t_i - s_i - \frac{2}{h}) n(V_i)(\alpha_i - 1)} P\left(B \cap \left(\frac{j-1}{h} \le \sigma < \frac{j}{h} \right) \right)$$

$$\longrightarrow P(B) \prod_i e^{(t_i - s_i) n(V_i)(\alpha_i - 1)} \quad (h \to \infty)$$

$$= P(B) E\left[\prod_i \alpha_i^{N(X, [s_i, t_i) \times V_i)} \right].$$

Thus the left side $\ge$ the right side in (1.6). Similarly the opposite inequality holds. This completes the proof. $\square$

Remark Notice that the strong renewal property does not hold for $M \in \mathscr{P}$ in general. For example, take V with $n(V) < \infty$. Then the time points at which X is in V form a sequence tending to ∞ a.s. Let σ be the first point in this sequence. Then

$$0 \in D_{\theta_\sigma X} \quad \text{and} \quad \theta_\sigma X(0) = X(\sigma) \in V \quad \text{a.s.},$$

while

$$0 \notin D_X \quad \text{a.s.}$$

References

1. Itô, K.: Stochastic processes. In: Barndorff-Nielsen, O.E., Sato, K. (eds.) Lectures given at Aahus University, Springer (2004)
2. Lévy, P.: Théorie de l'addition des variables aléatoires (2nd ed. 1954). Paris (1937)

Chapter 2
Application to Markov Processes

2.1 Problem

Let X_t be a standard Markov process with the state space S. The time interval $[0, \infty)$ is denoted by T. Let a be a fixed state and σ_a the hitting time for a. We impose the following four assumptions.

(A-1) $P_b(\sigma_a < \infty) = 1$;
(A-2) $E_b(\sigma_a \wedge 1) \to 0$ as $b \to a$;
(A-3) $\inf_{b \in U^c} E_b(\sigma_a \wedge 1) > 0$ for every neighborhood U of a;
(A-4) a is a *discontinuous exit state*.

We will explain the meaning of this condition.
s is called an *exit time* from a for the path $(X_t(\omega))$ if, for every $\varepsilon > 0$,

$$\{t : X_t(\omega) = a\} \cap (s - \varepsilon, s) \neq \emptyset$$

and if, for some $\varepsilon > 0$,

$$\{t : X_t(\omega) = a\} \cap (s, s + \varepsilon) = \emptyset.$$

All exit times from a for the path $(X_t(\omega))$ form a countable set depending on ω.

An exit time s from a for the path $(X_t(\omega))$ is called a *continuous* or *discontinuous* exit time according as

$$X_s(\omega) = a \quad \text{or} \quad X_s(\omega) \neq a.$$

a is called a *discontinuous exit state* if all exit times from a for the path $(X_t(\omega))$ are discontinuous a.s. with respect to P_a.

© The Author(s) 2015
K. Itô, *Poisson Point Processes and Their Application to Markov Processes*,
SpringerBriefs in Probability and Mathematical Statistics,
DOI 10.1007/978-981-10-0272-4_2

Let $X_t^0 = X_{t \wedge \sigma_a}$. Since the hitting time σ_a^0 of the path $(X_t^0(\omega))$ is the same as σ_a, the conditions (A-1), (A-2) and (A-3) are equivalent to

(A^0-1) $P_b(\sigma_a^0 < \infty) = 1$;
(A^0-2) $E_b(\sigma_a^0 \wedge 1) \to 0$, as $b \to a$;
(A^0-3) $\inf_{b \in U^c} E_b(\sigma_a^0 \wedge 1) > 0$ for every neighborhood U of a.

By the strong Markov property of (X_t), the probability laws of the path (X_t) is determined by the probability laws of the path (X_t^0) and the probability law of the path (X_t) starting at a. Symbolically we have

$$\text{p.l. of } (X_t) = \text{ p.l. of } (X_t^0) + \text{ p.l. of } (X_t) \text{ starting at } a. \tag{2.1}$$

Since the path (X_t) starting at a behaves outside of a in the same way as the path (X_t^0), the union relation in (2.1) is no disjoint union. We want to extract some information I from the probability law of the path (X_t) starting at a to obtain a symbolic information relation

$$\text{p.l. of}(X_t) = \text{p.l. of}(X_t^0) + I \quad \text{(disjoint union)}. \tag{2.1'}$$

In the subsequent sections we will prove that I consists of two elements: *jumping-in measure $k(db)$* and *stagnancy rate m*.

2.2 The Poisson Point Process Attached to a Markov Process at a State a

We use the same notations as in Sect. 2.1 and impose the conditions (A-1), (A-2), (A-3) and (A-4).

Let $A(t)$ be a local time of (X_t) at a. By our assumptions (A-1) and (A-2), $A(t)$ is determined up to a multiplicative constant and we have

$$P_b(A(t) < \infty \text{ for every } t) = 1$$

and

$$P_b(A(t) \to \infty \text{ as } t \to \infty) = 1.$$

We refer the reader to Blumenthal and Getoor [1] for the definition and the properties of local times.

Let U be the space of all right continuous functions: $T \to S$ with left limits. The sample path of (X_t) belongs to U a.s. for every starting point.

From now on we will refer to P_a for the probability law of $X_t(\omega)$ unless the contrary is explicitly stated. Let us define a point process $X : T \to U$ by

$$D_X = \{A(s) : s \text{ moves over all exit times from } a \text{ for the path}\},$$
$$X_\omega(t) = (X \circ \theta_{A^{-1}(t-)})^0 \quad \text{for} \quad t \in D_X,$$

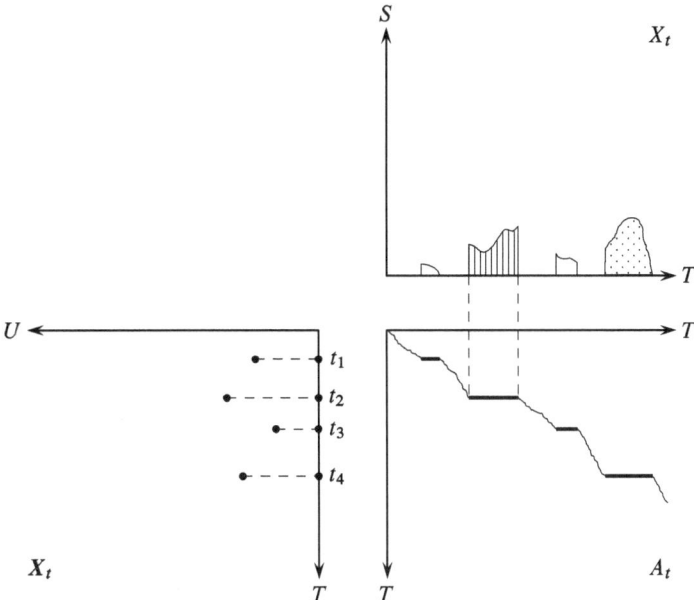

Fig. 2.1 X_t, A_t and X_t

where θ_t is a shift operator and 0 is a stopping operator. Note that D_X consists of all values of $A(t)$ corresponding to the flat t-intervals of $A(t)$ and that $X_\omega(t)$ is a function : $T \to S$ belonging to U for ω and t fixed and (Fig. 2.1)

$$X_\omega(t)(s) = X_{s \wedge \sigma_a(\theta_\tau \omega)}(\theta_\tau \omega) \quad \text{for} \quad s \in T, \quad \tau = A^{-1}(t-).$$

Figure 2.2 is an intuitive picture of X_t.

Fig. 2.2 An intuitive picture of X_t

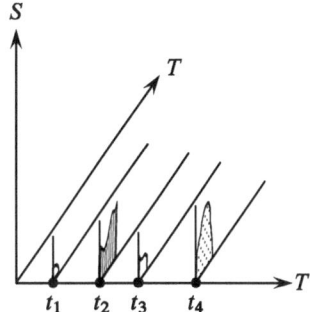

Theorem and Definition 2.2.1 *The point process X defined above is a Poisson point process: $T \to U$; it is called* the Poisson point process attached to the Markov process (X_t).

Proof Let $\{\mathscr{B}_t\}$ be a family of sub-σ-algebras of $\mathscr{B}$ in the definition of the Markov process (X_t). Since $A^{-1}(t+)$ and $A^{-1}(t-) = \sup_n A^{-1}(t - \frac{1}{n}+)$ are both stopping times, $\mathscr{B}_{A^{-1}(t-)}$ and $\mathscr{B}_{A^{-1}(t+)}$ are well-defined.

Let $\mathscr{B}_t(X)$ be the σ-algebra generated by $X|_d[0, t)$. Then

$$\mathscr{B}_t(X) \subset \mathscr{B}_{A^{-1}(t-)} \subset \mathscr{B}_{A^{-1}(t+)}.$$

For t fixed, we have

$$P(X(A^{-1}(t+)) = a) = 1.$$

Thus, for $B \in \mathscr{B}_t(X)$ and $M \in \mathscr{P}$, we have

$$P_a(B \cap ((\theta_t X) \in M)) = P_a(B)P_a(X \in M)$$

because of the additivity of $A(t)$, the strong Markov property of (X_t) and the definition of X. Thus X has renewal property. This implies that X is differential and stationary.

To prove the σ-discreteness of X we will introduce a map $h : U \to T$ by

$$h(u) = \inf\{t : u(t) = a\}.$$

If u is the path of $(X_t(\omega))$, then $h(u)$ will be $\sigma_a(\omega)$.

Let U_n be the set of all u such that

$$h(u) > \frac{1}{n}.$$

By (A-4) we have

$$X = \bigvee_n X|_r U_n \quad \text{a.s.}$$

Since

$$N(X, [0, t) \times U_n) \le n \cdot A^{-1}(t-) < \infty \quad \text{a.s.},$$

$X_\omega|_r U_n$ is discrete a.s. for each n. X is therefore σ-discrete. $\qquad\square$

2.3   The Jumping-In Measure and the Stagnancy Rate

Let us consider a map $e : U \to S$ by

$$e(u) = u(0), \quad u \in U.$$

Since the path of X_t has no discontinuities of the second kind and since $A^{-1}(t) < \infty$ for $t < \infty$, the distance $\rho(X_t, a)$ between X_t and a can be larger than $\varepsilon(>0)$ a finite number of times during $[0, A^{-1}(t))$ a.s. for $t < \infty$. This implies that $e \cdot X$ is a σ-discrete point process. By Theorem 1.6.1, we see that $e \cdot X$ is a Poisson point process.

Definition 2.3.1   The characteristic measure k of $e \cdot X$ is called the *jumping-in measure* of the Markov process X_t from a.

It is obvious that $k = n_X e^{-1}$. Since n_X is concentrated on the paths starting at points in $S - \{a\}$ by (A-4), k is concentrated on $S - \{a\}$. It is obvious that the total measure of k is the same as that of n_X. Since X is σ-discrete, the total measure of k is σ-finite.

Since $A^{-1}(t)$ is known to be an increasing homogeneous Lévy process ($=$a subordinator), it can be written as

$$A^{-1}(t) = m \cdot t + J(t), \quad m > 0,$$

when $J(t)$ is a pure jump process.

Definition 2.3.2   The coefficient m is called the *stagnancy rate* of the Markov process (X_t).

The following theorem shows that the characteristic measure n_X is determined by the measure k and the probability law of the path of (X_t^0).

Theorem 2.3.3

$$n_X(V) - \int_S k(db) P_b(X^0 \in V)$$

where X^0 denotes the path $(X_t^0(\omega), t \in T)$.

Proof Let S_i denote the set $\{b \in S : \rho(a, b) > 1/i\}$ for $i = 1, 2, \ldots$. Then $\cup S_i = S - \{a\}$. Let $U_i = \{u \in U : u(0) \in S_i\} = e^{-1}(S_i)$. Then U_i increases with i and the limit U_∞ is the space of all paths in U starting from points in $S - \{a\}$. We have

$$X = X|_r U_\infty = \bigvee_i X_i, \quad X_i = X|_r U_i$$

by (A-4). The set $A^{-1}(D_{X_i}) \cap [0, A^{-1}(t+)]$ is included in the set of the time points $s \in [0, A^{-1}(t+)]$ for which $\rho(X_s, X_{s-}) > 1/i$. Since the sample path of (X_t) has no

discontinuity points of the second kind, the latter set is finite and so is $A^{-1}(D_{X_i}) \cap$
$[0, A^{-1}(t+)]$. This implies $D_{X_i} \cap [0, t]$ is finite. X_i is therefore a discrete Poisson
point process.

By Theorem 1.4.1, we have

$$n_{X_i}(V_i) = \lambda_i P_a(X_i(\tau_i) \in V_i), \quad V_i \in \mathcal{U}_i \equiv U_i \cap \mathcal{U},$$

where $\lambda_i = n_{X_i}(U_i)$ and τ_i is the smallest element in D_{X_i}. By the definition we have

$$X_i(\tau_i) = X(\tau_i) = (X \circ \theta_{\sigma_i})^0, \quad \sigma_i = A^{-1}(\tau_i-).$$

Since $n_{X_i} = n_X | U_i$ and since σ_i is a stopping time with respect to $\{\mathcal{B}_t\}$, we have, for
$V \in \mathcal{U}$,

$$n_X(U_i \cap V) = \lambda_i P_a((X \circ \theta_{\sigma_i})^0 \in V \cap U_i)$$
$$= \lambda_i \int_{S_i} P_a(X_{\sigma_i} \in db) P_b(X^0 \in V \cap U_i)$$

Set $V = e^{-1}(B_i)$, $B_i \in \mathscr{S}_i \equiv S_i \cap \mathscr{S}$. Then $V \subset e^{-1}(S_i) = U_i$ and so

$$k(B_i) = \lambda_i P_a(X_{\sigma_i} \in B_i).$$

Thus we have

$$n_X(U_i \cap V) = \int_{S_i} k(db) P_b(X^0 \in V \cap U_i).$$

Letting $i \uparrow \infty$, we have

$$n_X(V) = \int_{S-\{a\}} k(db) P_b(X^0 \in V) = \int_S k(db) P_b(X^0 \in V),$$

which completes the proof.   □

The jumping-in measure k is not arbitrary. We have:

Theorem 2.3.4 k *is concentrated on* $S - \{a\}$ *and*

$$\int_S E_b(\sigma_a^0 \wedge 1) k(db) < \infty.$$

Proof $h \cdot X$ is also a Poisson point process whose integrated process is the discontinuous part of the increasing homogeneous Lévy process $A^{-1}(t)$. Therefore $h \cdot X$ is summable and so

$$\int_0^\infty (1 \wedge t) n_{h \cdot X}(dt) < \infty$$

by virtue of Theorem 1.7.3. Since $n_{h \cdot X} = n_X h^{-1}$, this can be written

$$\int_0^\infty (1 \wedge t) \int_S k(db) P_b(\sigma_a^0 \in dt) < \infty$$

by the previous theorem, namely

$$\int_S k(db) E_b(\sigma_a^0 \wedge 1) < \infty.$$

☐

Remark By this theorem and the condition (A-3), we have

$$k(U^c) < \infty$$

for every neighborhood U of a.

If $k(S) < \infty$, then X is discrete. Then the set $\{t : X_t(\omega) = a\}$ is a sequence of disjoint intervals ordered linearly and $A(t, \omega)$ is the sojourn time at a singleton $\{a\}$ up to a multiplicative constant. Thus we have $m > 0$ in the decomposition:

$$A^{-1}(t) = mt + J(t),$$

$J(t)$ being the discontinuous part of $A^{-1}(t)$. Therefore we obtain:

Theorem 2.3.5 $m \geq 0$ in general, and $m > 0$ in case $k(S) < \infty$.

$A(t)$ is determined up to a multiplicative constant and m and k depend on which version of $A(t)$ we take. Let $A_i(t)$, $i = 1, 2$ be two versions of $A(t)$ and write the corresponding m and k as m_i and k_i, $i = 1, 2$. Then we have a constant $c > 0$ such that

$$A_2(t) = c A_1(t).$$

Consider the decompositions

$$A_i^{-1}(s) = m_i s + J_i(s), \quad i = 1, 2.$$

Then

$$A_2^{-1}(cs) = A_1^{-1}(s),$$
$$m_2 cs + J_2(cs) = m_1 s + J_1(s)$$

and so

$$m_2 = \frac{1}{c} m_1.$$

Writing $\#A$ for the number of points in A, we have

$$
\begin{aligned}
\varepsilon \cdot k_2(B) &= E_a[\#\{s : 0 \le s \le \varepsilon, \ e(X_s) \in B\}] \\
&= E_a[\#\{s : 0 \le s \le \varepsilon, \ X(A_2^{-1}(s-)) \in B\}] \\
&= E_a[\#\{t : 0 \le A_2(t) \le \varepsilon, \ X(t) \in B\}] \\
&= E_a[\#\{t : 0 \le cA_1(t) \le \varepsilon, \ X(t) \in B\}] \\
&= E_a\left[\#\left\{t : 0 \le A_1(t) \le \frac{1}{c}\varepsilon, \ X(t) \in B\right\}\right] \\
&= \frac{1}{c}\varepsilon k_1(B)
\end{aligned}
$$

and so

$$
k_2 = \frac{1}{c}k_1.
$$

Thus we have:

Theorem 2.3.6 *If $A_2(t) = cA_1(t)$, then $m_2 = \frac{1}{c}m_1$ and $k_2 = \frac{1}{c}k_1$.*

Therefore m and k are determined up to a common multiplicative constant.

To have m and k determined uniquely, we have to take a standard version of the local time $A(t)$.

Definition 2.3.7 $A(t)$ is called *standard* if

$$
E_a\left(\int_0^\infty e^{-t}\, dA(t)\right) = 1,
$$

in which case

$$
E_b\left(\int_0^\infty e^{-t}\, dA(t)\right) = E_b(e^{-\sigma_a^0}) \quad \text{for every } b.
$$

The m and k that correspond to the standard $A(t)$ are called the *standard stagnancy rate* and the *standard jumping-in measure*.

Theorem 2.3.8 *The standard stagnancy rate m and the stagnancy jumping-in measure k satisfy the following conditions.*

(a) $m \ge 0$ in general and $m > 0$ in case $k(S) < \infty$.
(b) k is concentrated on $S - \{a\}$ and

    (i) $\int_S k(db)E_b(\sigma_a^0 \wedge 1) < \infty$;
    (ii) $m + \int_S k(db)E_b(1 - e^{-\sigma_a^0}) = 1$.

Proof By Theorems 2.3.4 and 2.3.5 it is enough to prove (b)-(ii). Since m and k are standard, the corresponding $A(t)$ satisfies

$$E_a\left(\int_0^\infty e^{-t}\,dA(t)\right) = 1.$$

But the left side is

$$E_a\left(\int_0^\infty e^{-A^{-1}(t)}\,dt\right) = \int_0^\infty E_a(e^{-A^{-1}(t)}\,dt)$$

$$= \int_0^\infty e^{-mt-t\cdot\int_0^\infty(1-e^{-s})\int_S k(db)P_b(\sigma_a^0\in ds)}\,dt$$

(see the proof of Theorem 2.3.4 and use the Lévy–Khinchin formula)

$$= \left(m + \int_0^\infty (1-e^{-s})\int_S k(db)P_b(\sigma_a^0 \in ds)\right)^{-1}$$

$$= \left(m + \int_S k(db)E_b(1 - e^{-\sigma_a^0})\right)^{-1}.$$

This proves (ii).   □

2.4 The Existence and Uniqueness Theorem

Suppose that X_t is a standard Markov process with the state space S and that a is a fixed state. We assume (A-1), (A-2), (A-3) and (A-4) in Sect. 2.1.

Let $X_t^0 = X_{t\wedge\sigma_a}$, m the standard stagnancy rate and k the jumping-in measure for X_t. Then we have proved:

(i) X_t^0 is a standard Markov process which satisfies (A^0-1), (A^0-2), (A^0-3).
(ii) m and k satisfy (a) and (b) in Theorem 2.3.8.

Now we want to construct X_t for X_t^0, m and k given.

Theorem 2.4.1 *Suppose that X_t^0, m and k satisfy (i) and (ii). Then there exists a standard Markov process X_t satisfying (A-1), (A-2) and (A-3) such that $X_{t\wedge\sigma_a}$ is equivalent to X_t^0 and that the standard stagnancy rate and the standard jumping-in measure are respectively equal to m and k. Such X_t is unique up to equivalence.*

Proof of existence First we will construct the Poisson point process X attached to the Markov process X_t that is to be constructed.

Let U be the space of all right continuous functions: $T \to S$ with left limits. Define a σ-finite measure n on U by

$$n(V) = \int_S k(db)P_b(X^0 \in V)$$

and construct a Poisson point process $X : T \to U$ with $n_X = n$ by Theorem 1.3.5, or by Theorem 1.5.4.

Set

$$\tilde{A}(s) = ms + \sum_{\substack{\alpha \le s \\ \alpha \in D_X}} h(X(\alpha))$$

where $h(u) = \inf\{\alpha \in T : u(\alpha) = a\}$.

Define $Y(t)$ as follows.

$$Y(t) = \begin{cases} X(s)(t - \tilde{A}(s-)) & \text{if } \tilde{A}(s-) \le t < \tilde{A}(s) \\ a & \text{if } \tilde{A}(s-) = t = \tilde{A}(s). \end{cases}$$

Now define the probability law P_a of the path of X_t starting at a by

$$P_a(X_\bullet \in V) = P(Y(\cdot) \in V)$$

and the probability law P_b of the path of X_t starting at a general state b by

$$P_b(X_{\bullet \wedge \sigma_a} \in V_1, X \circ \theta_{\sigma_a} \in V_2) = P_b(X_\bullet^0 \in V_1) P_a(X_\bullet \in V_2).$$

It is needless to say that the definition of P_a is suggested by Fig. 2.1 and that the definition of P_b is suggested by the strong Markov property.

First we will prove that

$$P(\tilde{A}(s) < \infty \text{ for every } s \text{ and } \tilde{A}(\infty) = \infty) = 1, \tag{2.2}$$

so that $Y(t)$ is well-defined for every t. If $k(S) = 0$, then $m > 0$ and

$$\tilde{A}(s) = ms < \infty \quad \text{and} \quad A(\infty) = \infty.$$

If $k(S) > 0$, then $h \cdot X$ is a Poisson point process with

$$n_{h \cdot X} = nh^{-1} = \int_S k(db) P_b(\sigma_a^0 \in \cdot).$$

Since

$$\int_0^\infty (t \wedge 1) n_{h \cdot X}(dt) = \int_S k(db) E_b(\sigma_a^0 \wedge 1) < \infty,$$

$h \cdot X$ is summable and so

$$J(s) = \sum_{\substack{\alpha \le s \\ \alpha \in D_X}} h(X(\alpha)) < \infty$$

for every $s < \infty$ and $J(s)$ is a homogeneous Lévy process with increasing paths. Since $n_{h \cdot X}([0, \infty)) = k(S) > 0$, we have

$$P(J(\infty) = \infty) = 1.$$

This proves (2.2).

Now we will prove that the process X_t defined above is a standard Markov process with (A-1), (A-2), (A-3) and (A-4).

Case 1. $k(S) < \infty$. In this case we have $m > 0$.

Since

$$n_X(U) = \int_S k(db) P_b(X_\bullet^0 \in U) = k(S),$$

X is discrete.

Set

$$D_X = \{\tau_1 < \tau_1 + \tau_2 < \tau_1 + \tau_2 + \tau_3 < \cdots\}$$

and set

$$\xi_i = X(\tau_1 + \tau_2 + \cdots + \tau_i), \quad i = 1, 2, \ldots.$$

Then $\tau_1, \tau_2, \ldots, \xi_1, \xi_2, \ldots$ are independent and

$$P(\tau_i > t) = e^{-tk(S)},$$

$$P(\xi_i \in V) = \frac{1}{k(S)} \int_S k(db) P_b(X^0(\cdot) \in V), \quad V \in \mathcal{U}.$$

In other words the probability law of ξ_i is the probability law of the path of X^0 with the initial distribution $k(db)/k(S)$.

By the definition of $Y(t)$ we have $Y(t) = a$ for

$$m\tau_1 + h(\xi_1) + \cdots + m\tau_{i-1} + h(\xi_{j-1})$$
$$\leq t < m\tau_1 + h(\xi_1) + \cdots + m\tau_{i-1} + h(\xi_{i-1}) + m\tau_i$$

and

$$Y(t) = \xi_i(t - m\tau_1 - h(\xi_1) - \cdots - m\tau_{i-1} - h(\xi_{i-1}) - m\tau_i)$$

for

$$m\tau_1 + h(\xi_1) + \cdots + m\tau_{i-1} + h(\xi_{i-1}) + m\tau_i$$
$$\leq t < m\tau_1 + h(\xi_1) + \cdots + m\tau_i + h(\xi_i).$$

Since

$$P(m\tau_i > t) = P(\tau_i > t/m) = e^{-tk(S)/m},$$

$X(t)$ can be described as follows. If it starts at a, it stays at a for an exponential holding time with the parameter $= k(S)/m$, then jumps into db with probability

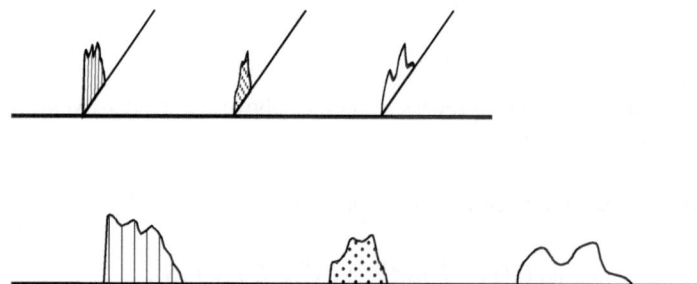

Fig. 2.3 Markov process from a point process

$k(db)/k(S)$ and moves in the same way as X_t^0 does until it hits a, it will repeat the same motion afterwards independently of its past history. If it starts at $b \neq a$, it performs the same motion as X_t^0 until it hits a and then it will act as above. We can verify the strong Markov property of this motion by routine. It is easy to check the other properties of X_t stated above (see Fig. 2.3).

Case 2. $k(S) = \infty$. Everything can be verified by routine except the fact that the sample path of $Y(t)$ belongs to U a.s. Since it is obvious that $Y(t)$ is right continuous and has left limits as far as it is in $S - \{a\}$, the only fact that needs proof is that the set of s such that

$$\sigma_\varepsilon(X(s)) < \infty, \quad \sigma_\varepsilon(u) = \inf\{t : \rho(a, u(t)) \geq \varepsilon\}$$

forms a discrete set a.s. for every $\varepsilon > 0$. Since $X(s)(t) = a$ for $t \geq h(X(s))$ a.s., $\sigma_\varepsilon(X(s)) < \infty$ is equivalent to

$$\sigma_\varepsilon(X(s)) < h(X(s))$$

a.s. It is therefore enough to prove that

$$X|_r V_\varepsilon, \quad V_\varepsilon = \{u : \sigma_\varepsilon(u) < h(u)\}$$

is discrete a.s., namely that

$$n_X(V_\varepsilon) < \infty.$$

Set

$$\delta = \inf\{E_b(\sigma_a^0 \wedge 1) : \rho(b, a) \geq \varepsilon\}.$$

Then $\delta > 0$ by (A^0-3).

Observe that

$$
\int_U h(u) \wedge 1 \, n_X(du) \geq \int_{V_\varepsilon} h(u) \wedge 1 \, n_X(du)
$$

$$
\geq \int_{V_\varepsilon} (h(u) - \sigma_\varepsilon(u)) \wedge 1 \, n_X(du)
$$

$$
= \int_V (h(u) - \sigma_\varepsilon(u)) \wedge 1 \int_S k(db) P_b(X^0 \in du)
$$

$$
= \int_S k(db) E_b[(\sigma_a^0 - \sigma_\varepsilon(X^0)) \wedge 1, \ \sigma_a^0 > \sigma_\varepsilon(X^0)]
$$

$$
= \int_S k(db) E_b[E_{X(\sigma_\varepsilon(X^0))}(\sigma_a^0 \wedge 1), \ \sigma_a^0 > \sigma_\varepsilon(X^0)]
$$

$$
\geq \delta \int_S k(db) P_b(\sigma_a^0 > \sigma_\varepsilon(X^0))
$$

$$
= \delta \int_S k(db) P_b(X^0 \in V_\varepsilon)
$$

$$
= \delta n_X(V_\varepsilon)
$$

and that

$$
\int_U h(u) \wedge 1 \, n_X(du) = \int_U h(u) \wedge 1 \int_S k(db) P_b(X^0 \in du)
$$

$$
= \int_S k(db) E_b(h(X^0) \wedge 1)
$$

$$
= \int_S k(db) E_b(\sigma_a^0 \wedge 1).
$$

Thus we have $n_X(V_\varepsilon) < \infty$.

The *proof of uniqueness* is easy, because the probability law of the path of (X_t^0) and k determine n_X and so the probability law of X, which, combined with m determines the probability law of the path of X_t. □

2.5 The Resolvent Operator and the Generator of the Markov Process Constructed in Sect. 2.4

The generator of a Markov process is defined in many ways which are not always equivalent to each other. We will adopt the following definition due to E.B. Dynkin.

Let X_t be a Markov process with right continuous paths. The *transition probability* $p(t, b, E)$ is defined by

$$
p(t, b, E) = P_b(X_t \in E),
$$

and the *transition operator* p_t is defined by

$$p_t f(b) = \int_S p(t, b, dc) f(c) = E_b(f(X_t)).$$

p_t carries the space $\mathbb{B}(S)$ of all bounded real Borel measurable functions into itself. It has the *semi-group property*:

$$p_{t+s} = p_t p_s, \quad p_0 = I \ (=\text{identity operator}).$$

The *resolvent operator* (*potential operator of order* α) R_α ($\alpha > 0$) is defined by

$$R_\alpha = \int_0^\infty e^{-\alpha t} p_t \, dt$$

i.e.,

$$R_\alpha f(b) = \int_0^\infty e^{-\alpha t} p_t f(b) \, dt = E_b \left(\int_0^\infty e^{-\alpha t} f(X_t) \, dt \right).$$

It satisfies the *resolvent equations*:

$$R_\alpha - R_\beta + (\alpha - \beta) R_\alpha R_\beta = 0.$$

The *Dynkin subspace* $\mathbb{L}$ of $\mathbb{B}(S)$ is defined by

$$\mathbb{L} = \{ f \in \mathbb{B}(S) : \lim_{t \downarrow 0} p_t f(b) = f(b) \text{ for every } b \}.$$

$\mathbb{L}$ is a linear subspace of $\mathbb{B}(S)$.

Because of the right continuity of the path of X_t we have

$$\mathbb{C}(S) \subset \mathbb{L} \subset \mathbb{B}(S),$$

$\mathbb{C}(S)$ being the space of all bounded continuous real functions on S.

It is easy to see that

$$p_t \mathbb{L} \subset \mathbb{L}, \quad R_\alpha \mathbb{L} \subset \mathbb{L}.$$

In view of this fact we will regard p_t and R_α as operators : $\mathbb{L} \to \mathbb{L}$, unless the contrary is stated explicitly.

By virtue of the resolvent equation $\mathcal{R} = R_\alpha \mathbb{L}$ is independent of α. $R_\alpha : \mathbb{L} \to \mathcal{R}$ is 1–1 and so R_α^{-1} is well-defined.

Definition 2.5.1 The generator $\mathcal{G}$ of (X_t) is defined by

$$\mathcal{D}(\mathcal{G}) = \left\{ f \in \mathbb{L} : \frac{1}{t}(p_t f - f) \text{ converges boundedly as } t \to 0 \right.$$
$$\left. \text{to a function } \in \mathbb{L} \right\}$$

and

$$\mathcal{G}f(b) = \lim_{t \downarrow 0} \frac{1}{t}(p_t f(b) - f(b)), \quad f \in \mathcal{D}(\mathcal{G}).$$

Theorem 2.5.2 $\mathcal{D}(\mathcal{G}) = \mathcal{R} = R_\alpha \mathbb{L}$, $\mathcal{G}f = \alpha f - R_\alpha^{-1} f$, $f \in \mathcal{D}(\mathcal{G})$.

Let X_t be a standard Markov process and a be a fixed state. We assume that (A-1), (A-2) and (A-3) are satisfied. Let $X_t^0 = X_{t \wedge \sigma_a}$. Then X_t^0 is also a standard Markov process with (A^0-1), (A^0-2) and (A^0-3). We will denote the transition operator, the resolvent operator and the generator of X_t respectively by p_t, R_α and $\mathcal{G}$ and the corresponding operators for X_t^0 are denoted by p_t^0, R_α^0 and $\mathcal{G}^0$.

Theorem 2.5.3 $\mathcal{D}(\mathcal{G}) \subset \mathcal{D}(\mathcal{G}^0)$ and

$$\mathcal{G}f(b) = \mathcal{G}^0 f(b), \quad b \neq a,$$
$$\mathcal{G}^0 f(a) = 0.$$

Proof If $f \in \mathcal{D}(\mathcal{G})$, then

$$f = R_\alpha g, \quad g \in \mathbb{L}.$$

By Dynkin's formula we have

$$f(b) = E_b \left(\int_0^\infty e^{-\alpha t} g(X_t) \, dt \right)$$
$$= E_b \left(\int_0^{\sigma_a} e^{-\alpha t} g(X_t) \, dt \right) + E_b(e^{-\alpha \sigma_a} f(X_{\sigma_a}))$$
$$= E_b \left(\int_0^{\sigma_a} e^{-\alpha t} g(X_t) \, dt \right) + E_b(e^{-\alpha \sigma_a}) f(a).$$

Set

$$g^0(b) = \begin{cases} g(b) & b \neq a, \\ \alpha R_\alpha g(a) = \alpha f(a) & b = a. \end{cases}$$

Then

$$R_\alpha^0 g^0(b) = E_b \left(\int_0^\infty e^{-\alpha t} g^0(X_t^0) \, dt \right)$$
$$= E_b \left(\int_0^{\sigma_a} e^{-\alpha t} g(X_t) \, dt \right) + E_b(e^{-\alpha \sigma_a}) R_\alpha^0 g^0(a).$$

Since

$$R_\alpha^0 g^0(a) = \int_0^\infty e^{-\alpha t} \alpha f(a)\, dt = f(a),$$

we have

$$f(b) = R_\alpha^0 g^0(b).$$

To complete the proof of $\mathscr{D}(\mathscr{G}) \subset \mathscr{D}(\mathscr{G}^0)$, we need only prove that g^0 belongs to the Dynkin space $\mathbb{L}^0$ of X_t^0. Since $X_t^0 = a$ for $t \geq \sigma_a$, we have

$$p_t^0 g^0(a) = g^0(a) \longrightarrow g^0(a) \quad \text{as } t \downarrow 0.$$

Suppose $b \neq a$. Then $P_b(\sigma_a > 0) = 1$ and so

$$\lim_{t \downarrow 0} P_b(\sigma_a \leq t) = 0.$$

Therefore

$$
\begin{aligned}
&|p_t^0 g^0(b) - g^0(b)| \\
=&|E_b(g^0(X_t^0) - g^0(b))| \\
=&|E_b(g(X_t), t < \sigma_a) + E_b(g^0(a), t \geq \sigma_a) - g(b)| \\
=&|E_b(g(X_t)) - E_b(g(X_t), t \geq \sigma_a) + E_b(g^0(a), t \geq \sigma_a) - g(b)| \\
=&|E_b(g(X_t)) - g(b)| + (\|g\| + |g^0(a)|) P_b(t \geq \sigma_a) \longrightarrow 0,
\end{aligned}
$$

where $\|g\| = \sup_{c \in S} |g(c)|$. Since

$$f = R_\alpha g = R_\alpha^0 g^0,$$

we have

$$\mathscr{G} f = \alpha f - g, \quad \mathscr{G}^0 f = \alpha f - g^0$$

and so

$$\mathscr{G}^0 f(b) = \mathscr{G} f(b) \quad \text{for } b \neq a$$

and

$$\mathscr{G}^0 f(a) = \alpha f(a) - g^0(a) = 0. \qquad \square$$

Let X_t be the Markov process constructed from X_t^0, m and k in Sect. 2.4. The resolvent and the generator for X_t are denoted respectively by R_α and $\mathscr{G}$ and the corresponding operators for X_t^0 are denoted respectively by R_α^0 and $\mathscr{G}^0$.

We will discuss the relation between $(R_\alpha, \mathcal{G})$ and $(R_\alpha^0, \mathcal{G}^0)$. Let us make three cases.

Case 1. $k(S) = 0$. In this trivial case a is a trap for X_t and (X_t) is equivalent to (X_t^0), so that

$$R_\alpha = R_\alpha^0 \quad \text{and} \quad \mathcal{G} = \mathcal{G}^0.$$

Case 2. $0 < k(S) < \infty$. ($m > 0$ in this case.) a is an exponential holding state with the rate $k(S)/m$.

Theorem 2.5.4 *If* $0 < k(S) < \infty$, *then*

$$R_\alpha g(b) = R_\alpha^0 g(b) + E_b(e^{-\alpha \sigma_a^0}) R_\alpha g(a) \quad \text{for } b \neq a; \tag{2.3}$$

$$R_\alpha g(a) = \frac{mg(a) + \displaystyle\int_S k(db) R_\alpha^0 g(b)}{\alpha m + \displaystyle\int_S k(db) E_b(1 - e^{-\alpha \sigma_a^0})}; \tag{2.4}$$

$$\mathcal{G} f(b) = \mathcal{G}^0 f(b) \quad \text{for } b \neq a; \tag{2.5}$$

$$m\mathcal{G} f(a) = \int_S k(db)(f(b) - f(a)). \tag{2.6}$$

Proof Equation (2.3) is obvious by Dynkin's formula.

To prove (2.4), set

$$f(a) = R_\alpha g(a) \quad \text{and} \quad f^0(a) = R_\alpha^0 g(a).$$

The Poisson point process X attached to (X_t) is discrete. Let σ be the first point in D_X and τ be the first exit time from a for (X_t). Let Y_t be the process derived from X in Sect. 2.4. By Dynkin's formula, we have

$$f(a) = F_a\left(\int_0^\infty e^{-\alpha t} g(X_t)\, dt\right)$$

$$= E_a\left(\int_0^\tau e^{-\alpha t} g(X_t)\, dt\right) + E_a(e^{-\alpha \tau} f(X_\tau))$$

$$= E\left(\int_0^{m\sigma} e^{-\alpha t} g(a)\, dt\right) + E[e^{-\alpha m\sigma} f(X_\sigma(0))]$$

$$= g(a) E\left[\frac{1 - e^{-\alpha m\sigma}}{\alpha}\right] + E[e^{-\alpha m\sigma}] E[f(X_\sigma(0))].$$

Observe

$$E(e^{-\alpha m \sigma}) = \int_0^\infty e^{-\alpha m t} e^{-t k(S)} k(S) \, dt$$

$$= \frac{k(S)}{\alpha m + k(S)}$$

and

$$E[f(X_\sigma(0))] = \frac{1}{k(S)} \int_S k(db) f(b).$$

Therefore we have

$$f(a) = \frac{mg(a) + \displaystyle\int_S k(db) f(b)}{\alpha m + k(S)}, \tag{2.7}$$

which, combined with (2.3), implies

$$f(a) = \frac{mg(a) + \displaystyle\int_S k(db) f^0(b) + \displaystyle\int_S k(db) E_b(e^{-\alpha \sigma_a^0}) f(a)}{\alpha m + k(S)}.$$

Solving this for $f(a)$, we have

$$f(a) = \frac{mg(a) + \displaystyle\int_S k(db) f^0(b)}{\alpha m + \displaystyle\int_S k(db) E_b(1 - e^{-\alpha \sigma_a^0})},$$

which proves (2.4). Equation (2.5) is obvious by Theorem 2.5.3.
It follows from (2.7) that

$$m(\alpha f(a) - g(a)) = \int_S k(db)(f(b) - f(a)),$$

which proves (2.6).   □

Case 3. $k(S) = \infty$. a is an instantaneous state for (X_t).

Theorem 2.5.5 *Theorem 2.5.4 holds also in case $k(S) = \infty$:*

$$\left(\int_S k(db)(f(b) - f(a)) \right) = \lim_{\varepsilon \downarrow 0} \int_{\rho(b,a) > \varepsilon} k(db)(f(b) - f(a))$$

with the following proviso. If $m > 0$, (2.4) holds for g with

$$\lim_{b \to a} g(b) = g(a) \tag{2.8}$$

and (2.6) holds for $f = R_\alpha g$ with g satisfying the same condition.

Proof (2.3) and (2.5) are obvious. Let $\varepsilon > 0$ and set

$$S^{1,\varepsilon} = \{b \in S : \rho(b, a) \geq \varepsilon\},$$
$$S^{2,\varepsilon} = S - S^{1,\varepsilon},$$
$$U^{i,\varepsilon} = \{u \in U : u(0) \in S^{i,\varepsilon}\}, \quad i = 1, 2,$$
$$X^{i,\varepsilon} = X|_r U^{i,\varepsilon}, \quad i = 1, 2.$$

Let $Y^{2,\varepsilon}(t)$ be the process derived from $X^{2,\varepsilon}$ in the same way as Y_t was derived from X in Sect. 2.4. Since we fix ε for the moment, we omit ε in $S^{i,\varepsilon}$, $U^{i,\varepsilon}$ etc.
  Let

$$J(t, X) = \sum_{\substack{s \leq t \\ s \in D_X}} h(X_s).$$

Similarly for $J(t, X^i)$. X^1 is discrete. Let σ be the first element in D_{X^1}.
  Noticing that

$$s \in D_X, \quad s < \sigma \Longrightarrow s \in D_{X^2},$$

we have

$$J(\sigma -, X) = J(\sigma -, X^2), \quad X_\sigma = X_\sigma^1$$

and

$$Y_t = Y_t^2 \quad \text{for} \quad t < m\sigma + J(\sigma -, X) = m\sigma + J(\sigma -, X^2).$$

$$f(a) \equiv R_\alpha g(a)$$
$$= E\left(\int_0^\infty e^{-\alpha t} g(Y_t)\, dt\right)$$
$$= E\left(\int_0^{m\sigma + J(\sigma -, X)} e^{-\alpha t} g(Y_t)\, dt\right)$$
$$+ E\left[e^{-\alpha m\sigma - \alpha I(\sigma -, X)} \int_0^{h(X_\sigma)} e^{-\alpha t} g(X_\sigma(t))\, dt\right]$$
$$+ E\left[e^{-\alpha m\sigma - \alpha J(\sigma -, X) - \alpha h(X_\sigma)} \int_0^\infty e^{-\alpha t} g(Y(t, \theta_\sigma X|_d(0, \infty)))\, dt\right]$$
$$= E\left(\int_0^{m\sigma + J(\sigma -, X^2)} e^{-\alpha t} g(Y_t^2)\, dt\right)$$
$$+ E\left[e^{-\alpha m\sigma - \alpha J(\sigma -, X^2)} \int_0^{h(X_\sigma^1)} e^{-\alpha t} g(X_\sigma^1(t))\, dt\right]$$
$$+ E\left[e^{-\alpha m\sigma - \alpha J(\sigma -, X^2) - \alpha h(X_\sigma^1)} \int_0^\infty e^{-\alpha t} g(Y(t, \theta_\sigma X|_d(0, \infty)))\, dt\right]$$
$$= I_1 + I_2 + I_3.$$

X^1 and X^2 are independent. σ and X^1_σ are $\mathscr{B}(X^1)$ measurable and independent of each other. X^2, σ and X^1_σ are therefore independent of each other. Thus we have

$$I_2 = E[e^{-\alpha m\sigma - \alpha J(\sigma-, X^2)}]E\left[\int_0^{h(X^1_\sigma)} e^{-\alpha t} g(X^1_\sigma(t))\, dt\right]$$

$$= \int_0^\infty P(\sigma \in dt)e^{-\alpha mt} E[^{-\alpha J(t-, X^2)}]\int_{S^1}\frac{k(db)}{k(S^1)}E_b\left(\int_0^{\sigma_a^0} e^{-\alpha t} g(X^0_t)\, dt\right).$$

Since $J(t, X^2)$ is a Lévy process increasing with jumps whose Lévy measure is equal to

$$n_{h\cdot X^2}(dt) = \int_{S^2} k(db)P_b(\sigma_a^0 \in dt),$$

we have

$$E[e^{-\alpha J(t-, X^2)}] = e^{-t\int_0^\infty (1-e^{-\alpha s})n_{h\cdot X^2}(ds)}$$

$$= e^{-t\int_{S^2} k(db)E_b(1-e^{-\alpha\sigma_a^0})}.$$

It is obvious that

$$P(\sigma \in dt) = e^{-k(S^1)t} k(S^1)\, dt.$$

Therefore

$$I_2 = \frac{\displaystyle\int_{S^1} k(db)E_b\left(\int_0^{\sigma_a^0} e^{-\alpha t} g(X^0_t)\, dt\right)}{\alpha m + k(S^1) + \displaystyle\int_{S^2} k(db)E_b(1 - e^{-\alpha\sigma_a^0})}.$$

By the strong renewal property of X we have

$$I_3 = E[e^{-\alpha m\sigma - \alpha J(\sigma-, X^2) - \alpha h(X^1_\sigma)}]E\left[\int_0^\infty e^{-\alpha t} g(Y_t)\, dt\right]$$

$$= E[e^{-\alpha m\sigma - \alpha J(\sigma-, X^2)}]E[e^{-\alpha h(X^1_\sigma)}]f(a)$$

$$= \frac{\displaystyle f(a)\int_{S^1} k(db)E_b(e^{-\alpha\sigma_a^0})}{\alpha m + k(S^1) + \displaystyle\int_{S^2} k(db)E_b(1 - e^{-\alpha\sigma_a^0})}.$$

Thus we have

$$f(a) = I_1 + \frac{\displaystyle\int_{S^1} k(db)\left[E_b\left(\int_0^{\sigma_a^0} e^{-\alpha t} g(X^0_t)\, dt\right) + E_b(e^{-\alpha\sigma_a^0})f(a)\right]}{\alpha m + k(S^1) + \displaystyle\int_{S^2} k(db)E_b(1 - e^{-\alpha\sigma_a^0})}. \qquad (2.9)$$

To evaluate I_1, consider

$$f^2(a) \equiv E\left(\int_0^\infty e^{-\alpha t} g(Y_t^2)\, dt\right)$$

$$= I_1 + E\left(e^{-\alpha m\sigma - \alpha J(\sigma -, X^2)} \int_0^\infty e^{-\alpha t} g(Y(t, \theta_\sigma X^2|_d(0, \infty)))\, dt\right)$$

$$= I_1 + \int_0^\infty P(\sigma \in ds)$$

$$\times E\left(e^{-\alpha ms - \alpha J(s-, X^2)} \int_0^\infty e^{-\alpha t} g(Y(t, \theta_s X^2|_d(0, \infty)))\, dt\right)$$

$$= I_1 + \int_0^\infty P(\sigma \in ds)$$

$$\times E(e^{-\alpha ms - \alpha J(s-, X^2)}) E\left(\int_0^\infty e^{-\alpha t} g(Y(t, X^2))\, dt\right)$$

(by the renewal property of X^2)

$$= I_1 + \int_0^\infty P(\sigma \in ds) E(e^{-\alpha ms - \alpha J(s-, X^2)}) f^2(a)$$

$$= I_1 + E(e^{-\alpha m\sigma - \alpha J(\sigma -, X^2)}) f^2(a).$$

This implies

$$I_1 = f^2(a)[1 - E(e^{-\alpha m\sigma - \alpha J(\sigma -, X^2)})]$$

$$= f^2(a)\left[1 - \frac{k(S^1)}{\alpha m + k(S^1) + \int_{S^2} k(db) E_b(1 - e^{-\alpha \sigma_a^0})}\right]$$

$$= f^2(a)\frac{\alpha m + \int_{S^2} k(db) E_b(1 - e^{-\alpha \sigma_a^0})}{\alpha m + k(S^1) + \int_{S^2} k(db) E_b(1 - e^{-\alpha \sigma_a^0})}.$$

From (2.9) we have

$$f(a) = f^2(a)\frac{\alpha m + \int_{S^2} k(db) E_b(1 - e^{-\alpha \sigma_a^0})}{\alpha m + k(S^1) + \int_{S^2} k(db) E_b(1 - e^{-\alpha \sigma_a^0})} \tag{2.10}$$

$$+ \frac{\int_{S^1} k(db)\left[E_b\left(\int_0^{\sigma_a^0} e^{-\alpha t} g(X_t^0)\, dt\right) + E_b(e^{-\alpha \sigma_a^0}) f(a)\right]}{\alpha m + k(S^1) + \int_{S^2} k(db) E_b(1 - e^{-\alpha \sigma_a^0})}.$$

Solving this for f we have

$$f(a) = \frac{f^2(a)\left(\alpha m + \int_{S^2} k(db)E_b(1 - e^{-\alpha\sigma_a^0})\right) + \int_{S^1} k(db)E_b\left(\int_0^{\sigma_a^0} e^{-\alpha t}g(X_t^0)\,dt\right)}{\alpha m + \int_S k(db)E_b(1 - e^{-\alpha\sigma_a^0})}.$$

(2.11)

Let $\varepsilon \downarrow 0$, then

$$\int_{S^1} k(db)E_b\left(\int_0^{\sigma_a^0} e^{-\alpha t}g(X_t^0)\,dt\right) \longrightarrow \int_S k(db)E_b\left(\int_0^{\sigma_a^0} e^{-\alpha t}g(X_t^0)\,dt\right);$$

notice that

$$\int_S k(db)\left|E_b\left(\int_0^{\sigma_a^0} e^{-\alpha t}g(X_t^0)\,dt\right)\right|$$

$$\leq \|g\| \int_S k(db)E_b(1 - e^{-\alpha\sigma_a^0})$$

$$< \infty, \quad \| \ \| = \text{sup. norm}$$

by virtue of $\int_S k(db)E_b(\sigma_a^0 \wedge 1) < \infty$. It is obvious that

$$\int_{S^2} k(db)E_b(1 - e^{-\alpha\sigma_a^0}) \longrightarrow 0 \quad \text{as} \quad \varepsilon \downarrow 0.$$

It follows from (2.10) and (2.3) that

$$f(a) = f^2(a)\frac{\alpha m + \int_{S^2} k(db)E_b(1 - e^{-\alpha\sigma_a^0})}{\alpha m + k(S^1) + \int_{S^2} k(db)E_b(1 - e^{-\alpha\sigma_a^0})}$$

$$+ \frac{\int_{S^1} k(db)f(b)}{\alpha m + k(S^1) + \int_{S^2} k(db)E_b(1 - e^{-\alpha\sigma_a^0})}$$

so that

$$m(\alpha f(a) - \alpha f^2(a)) + f(a)\int_{S^2} k(db)E_b(1 - e^{-\alpha\sigma_a^0})$$

$$= f^2(a)\int_{S^2} k(db)E_b(1 - e^{-\alpha\sigma_a^0}) + \int_{S^1} k(db)(f(b) - f(a)).$$

(2.12)

If $m = 0$, we can derive (2.4) and (2.6) from (2.11) and (2.12), letting $\varepsilon \downarrow 0$ and noticing that

$$|f^2(a)| \equiv |f^{2,\varepsilon}(a)| \le \|g\|/\alpha.$$

If $m > 0$, we need only prove that

$$\lim_{\varepsilon \downarrow 0} f^{2,\varepsilon}(a) = \frac{g(a)}{\alpha}, \tag{2.13}$$

in order to derive (2.4) and (2.6) from (2.11) and (2.12).

Let $\eta > 0$ and set

$$V^1 = V^{1,\eta} = \{u \in U : \sup_t \rho(u(t), a) \ge \eta\},$$

$$V^2 = V^{2,\eta} = U - V^1,$$

$$Y^i = Y^{i,\varepsilon,\eta} = X^{2,\varepsilon}|_r V^{i,\eta}, \quad i = 1, 2.$$

By the argument in the last step of the existence proof of Theorem 2.4.1, we have

$$\lambda \equiv \lambda_{\varepsilon,\eta} \equiv n_{Y^1}(V^1) = n_{X^{2,\varepsilon}}(V^{1,\eta})$$

$$\le \left(\inf_{\rho(b,a) > \eta} E_b(\sigma_a^0 \wedge 1) \right)^{-1} \int_{S^{2,\varepsilon}} k(db) E_b(\sigma_a^0 \wedge 1)$$

$$\longrightarrow 0, \quad \varepsilon \downarrow 0 \quad \text{for } \eta \text{ fixed.}$$

Y^1 is a discrete Poisson point process. Let $\tau = \tau_{\varepsilon,\eta}$ be the first element in D_{Y^1}. Then τ is exponentially distributed with rate $= \lambda_{\varepsilon,\eta}$. Using the same argument as in deriving (2.9), we obtain

$$\left| f^{2,\varepsilon}(a) - \frac{g(a)}{\alpha} \right|$$

$$\le E\left(\int_0^\infty e^{-\alpha t} g_0(Y_t^{2,\varepsilon}) \, dt \right), \quad g_0(b) = |g(b) - g(a)|$$

$$= E\left(\int_0^{m\tau + J(\tau, Y^2)} e^{-\alpha t} g_0(Y(t, Y^2)) \, dt \right)$$

$$+ E(e^{-\alpha m\tau - \alpha J(\tau-, Y^2)}) E\left(\int_0^{h(Y_\tau^1)} e^{-\alpha t} g_0(Y_\tau^1(t)) \, dt \right)$$

$$+ E(e^{-\alpha m\tau - \alpha J(\tau-, Y^2) - \alpha h(Y_\tau^1)}) E\left(\int_0^\infty e^{-\alpha t} g_0(Y_t^{2,\varepsilon}) \, dt \right).$$

Since $\rho(Y(t, Y^2), a) < \eta$ for $0 < t < m\tau + J(\tau-, Y^2)$, we have

$$\left| f^{2,\varepsilon}(a) - \frac{g(a)}{\alpha} \right| \le \delta(\eta) \frac{1}{\alpha} + E(e^{-\alpha m\tau}) \frac{\|g_0\|}{\alpha} + E(e^{-\alpha m\tau}) \frac{\|g_0\|}{\alpha}$$

where $\delta(\eta) = \sup\{g_0(b), \rho(b, a) < \eta\} \to 0$ $(\eta \downarrow 0)$ by (2.8). Since τ is exponentially distributed with rate $\lambda_{\varepsilon,\eta}$, we have

$$E(e^{-\alpha m \tau}) = \int_0^\infty e^{-\alpha m t} e^{-\lambda_{\varepsilon,\eta} t} \lambda_{\varepsilon,\eta} \, dt = \frac{\lambda_{\varepsilon,\eta}}{\alpha m + \lambda_{\varepsilon,\eta}}$$

$$\longrightarrow 0, \quad \varepsilon \downarrow 0$$

by $m > 0$. Thus we have

$$\limsup_{\varepsilon \downarrow 0} \left| f^{2,\varepsilon}(a) - \frac{g(a)}{\alpha} \right| \le \delta(\eta) \cdot \frac{1}{\alpha} \longrightarrow 0, \quad \eta \downarrow 0.$$

This completes the proof. $\square$

2.6   Examples

Example 1 Let $S = [0, \infty)$ and X^0 be a diffusion in S stopped at 0 such that the generator of X^0 is

$$\mathscr{G}^0 = \frac{d}{dm} \frac{d}{dx}.$$

Let 0 be an *exit or regular* (i.e., *exit* in Feller's new terminology) *boundary*, i.e.,

$$\int_0^1 m(\xi, 1) \, d\xi < \infty.$$

Then X^0 satisfies $(A^0\text{-}1)$, $(A^0\text{-}2)$ and $(A^0\text{-}3)$ in Sect. 2.1; notice that

$$\inf_{\rho(b,0) > \varepsilon} E_b(\sigma_0^0 \wedge 1) = E_\varepsilon(\sigma_0^0 \wedge 1) > 0.$$

We will investigate the condition (i) in Theorem 2.3.8:

$$\int_S k(db) E_b(\sigma_0^0 \wedge 1) < \infty. \tag{2.14}$$

This is equivalent to

$$\int_S k(db) E_b(1 - e^{-\sigma_0^0}) < \infty.$$

Since $u(b) = E_b(e^{-\sigma_0^0})$ is a decreasing positive solution of

$$\frac{d}{dm}\frac{d}{dx}u = u, \quad u(0) = 1,$$

$$u'(1) - u'(\xi) = \int_\xi^1 u(\xi)m(d\xi) \approx m(\xi, 1) \quad (\xi \downarrow 0),$$

$$u(0) - u(b) \approx \int_0^b (m(\xi, 1) - u'(1))\, d\xi$$

$(\alpha(\xi) \approx \beta(\xi)$ $(\xi \downarrow 0)$ means that we have $c_1, c_2 > 0$ independent of ξ such that $c_1\beta(\xi) < \alpha(\xi) < c_2\beta(\xi)$ near $\xi = 0)$.

Case 1. (regular case) If 0 is a *regular* (i.e., *exit and entrance* in Feller's new terminology) *boundary*, i.e., $m(0, 1) < \infty$, then

$$E_b(1 - e^{-\sigma_0^0}) = u(0) - u(b) \approx b \quad (b \downarrow 0).$$

Since $E_b(1 - e^{-\sigma_0^0}) \to 1$ as $b \to \infty$, $E_b(1 - e^{-\sigma_0^0}) \approx b \wedge 1$ in $0 < b < \infty$. Therefore our condition (2.14) turns out to be

$$\int_0^\infty k(db)(b \wedge 1) < \infty.$$

Case 2. (exit case) If 0 is an *exit* (i.e., *exit and non-entrance* in Feller's new terminology) *boundary*, i.e., $m(0, 1) = \infty$, then (2.14) turns out to be

$$\int_0^\infty k(db)\left[\int_0^b m(\xi, 1)\, d\xi \wedge 1\right] < \infty.$$

Example 2 Let $S = [0, \infty)$ and X^0 be a deterministic motion with constant speed "-1". Then $P_b(\sigma_0^0 = b) = 1$ and so (2.14) is written as

$$\int_0^\infty k(db)(b \wedge 1) < \infty.$$

Reference

1. Blumenthal, R., Getoor, R.: Markov Processes and Potential Theory. Academic Press, New York (1968)